AMELIA EARHART'S DAUGHTERS

AMELIA EARHART'S DAUGHTERS

The Wild and Glorious Story of American Women Aviators from World War II to the Dawn of the Space Age

LESLIE HAYNSWORTH
and
DAVID TOOMEY

Perennial
An Imprint of HarperCollins*Publishers*

A hardcover edition of this book was published in 1998 by William Morrow and Company, Inc.

HarperCollins books may be purchased for educational, business, or sales promotional use. For information please write: Special Markets Department, HarperCollins Publishers Inc., 10 East 53rd Street, New York, NY 10022.

First Perennial edition published 2000.

Designed by Oksana Kushnir

Library of Congress Cataloguing-in-Publication Data has been applied for.

ISBN 0-380-72984-9 (pbk.)

00 01 02 03 04 ❖/RRD 10 9 8 7 6 5 4 3 2 1

For our families

ACKNOWLEDGMENTS

The narrative which follows is based upon hundreds of individual sources, among them Air Force and NASA records, government transcripts, personal diaries, and published and unpublished nonfiction accounts. It is also derived from interviews with the women whose lives are depicted on these pages. Although we have taken pains to ensure that the text is accurate, memories fail, and our research has found that even official records are in many places contradictory. In places where accounts disagree, this work reflects the position of the majority, or that account which for a variety of reasons seems most credible. No names have been changed, and all excerpts from official and personal correspondence are verbatim. Although in some cases two or more conversations have been combined in the interests of avoiding unnecessary repetition, in all cases the dialogue represents the personalities and positions of their speakers as we understand them. Most dialogue is taken verbatim or nearly verbatim from that which was reported or reconstructed in diaries, transcripts, and/or interviews. Any errors of fact are ours alone.

We are grateful to Joseph D. Atkinson Jr. of Johnson Space Center's Equal Opportunity Office for insight into NASA's early struggles with questions of astronaut qualifications; to Lee Saegesser, Mark Kahn, and Steve Garber of the NASA History Office in Washington, D.C., for providing documents from the space program's early history;

to Jenna Kimberlin of the International Woman's Air and Space Museum, and Nancy Marshall Durr of Texas Woman's University Libraries for allowing us access to WASP photograph collections; to C. E. Ruckdaschel of Rockwell International for information concerning the X-15 Research Program, and to Pauline Vincent for sharing memories of her sister. For other background we are grateful to D. Menard of the United States Air Force Museum at Wright-Patterson Air Force Base; documentary film producer James Cross; Norman G. Richards and Tom Newman of the Archives Division of the Smithsonian Institution National Air and Space Museum; Darla Bullard of the Archives Division of the Resource Center of the Ninety-Nines, Inc.; and the Public Affairs Offices of Edwards Air Force Base, Patuxent Naval Test Center, and Johnson Space Center. For invaluable assistance in locating historical records, we thank the staffs of Alderman Library and the Science and Engineering Library at the University of Virginia, Newman Library at Virginia Tech, the Thomas Cooper Library at the University of South Carolina, the Richland County Public Library in Columbia, South Carolina, and Oakland Public Library in Oakland, California. The authors of the works whose titles appear in the Selected Bibliography at the end of this book provided us with a wealth of background material.

We also thank Heather Seagroatt for her friendship and inspiration, Mary Beth Oliver for technical assistance, Jim Dobberfuhl and Heather McGilvray for their hangar talk, Anne Shaffran and Jane Ward Shaffran for insights into military protocol, John Christman for his suggestion of a title, and Penelope Gleeson and her writing classes for help with a work-in-progress. We owe a special thanks to Carole Fungaroli for acting as navigator. We are indebted to Claire Wachtel for taking a chance on us, and to her editorial assistants, Tracy Quinn and Jessica Baumgardner. We are also indebted to David Hendin for his support, encouragement and unfailing good humor. We gratefully acknowledge the assistance of the English Department administrative staff at Virginia Tech, especially Evelyn Raines, Tanya Reece, and Nancy Stevens.

Of course, our greatest debt is to the women who are the subject of this book, and who graciously agreed to recite for us the stories they have already told many times over.

L.H.
D.M.T.

CONTENTS

PROLOGUE

Cape Canaveral—February 3, 1995, 12:10 A.M.

The seven women stand in the VIP section of the bleachers, a mile from the launchpad. It is after midnight, but the bleachers are brightly lit and the women's faces are clearly visible. The seven women are no longer young—a few are well into their sixties—but especially when they look in the same direction, something about them doesn't seem nearly that old. It's their eyes. The flesh around their eyes sags a little, but the eyes themselves are extraordinarily bright and clear. They could be the shining sleep-washed eyes of thirteen-year-old girls.

In 1960 and 1961, these women passed the same exhaustive medical tests given to the original seven Mercury astronauts. For a few months they had cause to believe that some of their number might fly into space. The reason their eyes are still extraordinarily bright and clear is simple. It's because they are the eyes of pilots.

The women are in the VIP section at Cape Canaveral to watch the launch of the Space Shuttle Discovery. For NASA, it is an especially important flight. It will be the first to rendezvous with the Russian space station Mir, and will be a necessary prelude to the construction of the International Space Station. But for these particular spectators the flight has another significance. It is the first time a United States spacecraft will be piloted by a woman. In fact, they are here at the behest of that woman—Air Force Lieutenant Colonel Eileen Collins. Colonel Collins is a maker of history who is also aware of it. She has brought aboard the Shuttle symbols of women's aviation past—a scarf that belonged to Amelia Earhart, the silver wings of the Women Airforce Service Pilots of World War II, and some trinkets from a few of the women in the bleachers. That she will be carrying some part of these women with her is a small but meaningful gesture, and in a sense it is Collins's attempt to repay them for their inspiration.

The seven women have been watching launches on television since the beginning of the Space Age—the tiny Redstones and Atlases, the larger and faster Titans, the gargantuan moon-bound Saturns, and finally the Shuttle itself. With the rest of the country, they were proud of Glenn's three orbits, and thrilled at the triumph of Armstrong's small step. And again with the rest of the country, they were gladdened by the flight of Sally Ride, then shocked at the loss of the Challenger. Like many Americans, they have grown accustomed to the rituals of countdown, the television images of the rockets rising slowly and then more quickly and then growing smaller and finally disappearing against a clear sky. They have listened to those countdowns and watched those images, after all, for thirty-seven years.

But actually to be here on this night is different—to feel the warm Gulf Stream breeze off the Banana River, to breathe the salt air, and to hear the announcements of launch preparations over the open-air loudspeaker system, to applaud and cheer when that loudspeaker announces that the crew is walking along the gantry platform to board the spacecraft, and mostly to see the thing itself, poised in the dazzling crossfire of searchlights, mist from its superchilled fuel tanks rolling down the sides. It's hard to believe that in a few minutes this enormous machine will climb into the sky on a column of fire.

The voice on the open-air loudspeaker announces that preparations are complete, and begins a countdown. The people standing there in the bleachers can't help but count, too. Before they can reach four, counting is beside the point, because the engines have ignited and smoke is growing and blooming all around the launchpad . . . now there's a steady rumble they can feel through their feet . . . and the clear fire roars from beneath, and then out of the solid rocket boosters on either side comes a fire so bright it hurts to look at. It's like a sunrise. Everything—the gantry, the billowing, mushrooming smoke, the bleachers, the palmetto grass—everything is suddenly lit with orange light. The huge machine is straining against the latches holding it to the gantry, and the steady rumble is louder, and latches holding it to the gantry break away, and unbelievably, impossibly, the enormous thing begins to lift itself. It's slowly rising . . . it clears the tower and somehow it's going fast and slow at the same time, and now the only connection to earth is the bright flames and smoke beneath it. People on the bleachers are saying "Go . . . go . . . go" and because they have to do *something* they start applauding although they can't hear anything over the deafening thunder of the engines. It's climbing faster and now they're just looking up at it, and it's going higher and it's moving faster and now it's high in the night sky. It begins a slow roll, and somehow this huge machine riding a pillar of fire way up in the middle of the dark air looks beautiful—there's no other word for it—and it's moving faster and faster and arcing out over the Atlantic. It's getting smaller, and it's harder to see the thing itself, but the fire beneath is as bright as a small sun now reflected in the water beneath, and now it's still getting smaller and it's so far out over the ocean they can see only the flame, and it looks like a tiny fierce white torch at the end of a long twisting trail of smoke. That torch splits into two torches like red embers, slowly falling. They fade and disappear, but at the place where they split there is still a bright star. It's reflected in the dark water below and the star grows fainter and fainter until it looks like the other stars near the eastern horizon, except it's moving. It's mostly dark around the launchpad again and there's a breeze coming off the river. It's quiet enough to hear the voices of the crew coming over the open-air loudspeaker. Through the static comes a woman's voice:

Houston—we are single engine press to MECO. And we are still go on all systems.

One of the seven women standing on the bleachers is Jerri Truhill. She is a smallish woman with a round face. Truhill is a retired pilot from Dallas. She is sixty-five now, but age seems only to have quickened her sense of humor and sharpened an already sharp tongue. Alone among the seven women, she has not taken up Collins's offer to carry anything—she had joked that this was out of a pilot's instinctive consideration for weight constraints. Truhill is here with her granddaughter Cindy, a young woman recently graduated from the University of Texas. When Discovery is just a point of light above the dark eastern horizon and they are still watching it, Cindy says, "Oh, Grandmother, I want to fly."

Truhill smiles, squeezes her granddaughter's hand a little, and says, "I've been waiting to hear that."

Colonel Collins's is the culmination of a long, much-interrupted story. It's a story of places that nobody has much heard of. Places like a backwoods settlement near a northern Florida sawmill—no more than a rough arrangement of shacks of scrap pine—named, with neither irony nor poetry, Millville. And a dusty, tumbleweed-haunted crossroads in west Texas, called Sweetwater. It's also a story of aircraft nobody much talks about anymore. Fast and dangerous machines like the P-38 Lightning and the P-26 Helldiver. Enormous beasts like the B-17 Flying Fortress and B-29 Superfortress. And strange and exotic earthbound spaceships like the YP-59A.

Mostly though, it's a story of women almost no one remembers. Some have hell-for-leather names like Jackie Cochran; some have plain, matter-of-fact names like Jerrie Cobb; and some have fine, soaring, lyrical names like Nancy Harkness Love. It's a story of women with unbelievable histories, histories in the air—of surviving crashes, of testing the first American jet, of flying low over Amazon jungles. And histories on the ground—making the case for women flyers in the corridors of the Pentagon, in the offices of NASA, and in the halls of the United States Congress.

For most, this story will be new. But it will also have a familiar resonance. It is a story of talented, resourceful and willing women, promises reneged upon and dreams deferred. In part, this is a story of heartbreak. But it is also a story of courage, tenacity and clear-eyed vision. Because for these women, the sky was not the limit. For these women there was no limit.

AMELIA EARHART'S DAUGHTERS

PART I

I

JACKIE

The man is not the kind who'd usually attract much notice. He has a faded look about him, the look of a man who has accepted that it is his lot in life to work hard for little reward. And around here, pretty much everyone has that look, which makes the faded man all the more unremarkable. "Here" is Millville, Florida, circa 1913. Millville is little more than a few rows of frail, windowless pine and tar-paper shacks, and most of the people who live in the shacks are transients. Work in these parts is seasonal, and when jobs come to an end, people have to move on. There is usually just about enough money to keep a family fed and clothed, but there is never any left over to put aside for the future. Consequently, Millville makes it very hard for its inhabitants to imagine a future that might be different and better than the life they know here. This is why the faded man is so difficult to distinguish from all the other faded men in Millville. Millville breeds faded men.

But today, in the main room of a shabby general store, this man is important. Today he has an audience, watching him with rapt attention. In his hand he holds a cardboard box, and in the box are slips of paper, and on the slips of paper are written the names of just about every young girl in Millville. Soon the man will reach into the box and grope around until his callused fingers come to rest on what feels like the right slip of paper. He will pull this paper out of the box and unfold it and hold it close to his eyes and squint at it and maybe mouth the name that's printed on it silently once or twice before he speaks it aloud. The girls are looking at him as if he matters. And to them he does. It is his hand that will choose the slip, his mouth that will pronounce the name of the girl who will win the doll.

To a different group of girls, the doll might look as ordinary as the faded man usually looks to the people of Millville. She is pretty, of course, with golden curls and big brown eyes and a sweet rosebud mouth. She is the most stereotypical of dolls, but to the girls of Millville she is exotic, and even a little unreal. To them, she is like a bright promise of a better world that Millville usually won't let them believe in. She just appeared in the commissary shop at the mill one day, and the girls would come in little clusters to gape at her silently. Then they learned the really amazing thing about her—that she was the prize in a contest. Any of them could win her. All they had to do to enter the contest was spend twenty-five cents on toys in the commissary. Each time they did this, their names would go into the box.

So the girls had to find quarters. And one of these girls decided that, no matter how much anyone else wanted the doll, she wanted it the most. But she figured that the strength of her wanting was not in itself enough. Wanting wouldn't put slips of paper in the box. Only money could do that. Money wasn't something she could ask her family for, because it wasn't something they ever had. She would have to earn her chance to own the doll. So she went looking for work, hiring herself out to anyone who would promise to pay her to do the kinds of jobs an eight-year-old girl can do. She hauled water out of the well until her hands cracked and bled, and she changed diapers and washed dishes and swept floors and minded other children. After three weeks her small body was profoundly tired, but all of this seemed to her a reasonable trade for even the possibility of having the doll. By her calculations, her labors had netted her at least three dollars. But when the time came to collect, the neighbors who hired her shrugged and looked sheepish and mumbled something about how they'd pay her soon. She knew from the way they said it that soon meant never. She ended up with fifty cents. Her name went into the box exactly twice.

The man stands at the front of the room with the box in his hand and the doll by his side and he looks at the rows of waiting girls. Finally he reaches in and slowly pulls a slip of paper out of the box. The girls hold their breath and lean forward a little, and the man squints at the paper and moves his mouth silently a few times. At last he comes out with it: The name on the paper is Jackie.

Later she will be Jackie Cochran, but Cochran is a name she will choose for herself, because she never knew what name she would have had if she'd been raised by her real parents. In Millville, she is just Jackie, a foster child, a tacked-on addition to a family of impoverished migrant workers. Like a lot of the girls here, she wears a dress made of burlap, and when the family can't feed her she survives by eating pine nuts she finds on the ground. She does not even know the story of how she came to be taken in by this family, and maybe it has never even occurred to her to ask. Millville does not encourage its inhabitants to wonder. But to Jackie, the doll virtually embodies wonder. The faded man presses the doll into her arms and she looks at the doll and wonders if it is after all possible that wanting can be enough.

Looking at the doll, little Jackie does not at first notice that her foster family has crowded around her, and she doesn't hear what they are saying, although the words are addressed to her. She is at first only vaguely annoyed by this relentless cacophony of voices, but the voices persist, and gradually it sinks in that they're trying to tell her something about her older foster sister, Mamie. She realizes that it's not Mamie herself they're talking about, it's Mamie's little daughter Willie Mae. Willie Mae is two.

"Doesn't Willie Mae deserve the doll for Christmas, Jackie?" It isn't so much a question as an order. "You're too old to play with dolls now, aren't you?"

Mutely, she shakes her head. She's never had a doll of her own. But the look in her eyes, at once mutinous, fierce and pleading, does nothing to alter her foster parents' will. "Give the doll to Willie Mae."

"I won't." Jackie says defiantly. "It's mine. I won it."

"You will."

The doll is taken from her arms and handed to little Willie Mae.

It took her almost twenty years, but Jackie Cochran got her doll back. She was living in Manhattan by then, working as a beautician at Saks Fifth Avenue. She'd come a long way from Millville, and she didn't find the past much worth remembering. But one day she got a phone call from Willie Mae, and Willie Mae said she had a little girl of her own now. She was flat broke, and she was wondering if maybe Aunt Jackie . . .

Jackie Cochran was willing to offer Willie Mae and her daughter a second chance, a new life in New York made possible with her financial support—but on one condition.

She wanted her doll back.

She got it.

For years Jackie was the only name she had. As a transient, she was out of school more than she was in it, so that by her teens she had achieved the equivalent of a second-grade education. She wasn't even entirely certain of her age, but when she was about ten or twelve, she and her foster family moved to Columbus, Georgia, and there, at last, they stayed. Along with the rest of the family, Jackie went to work at a cotton mill, and for a while life seemed to be looking up. There was food on the table and even money left over sometimes for an occasional trip to the shops, but when the mill workers went on strike, they were all laid off, and that appeared to be the end of the good life.

Jackie took a job, sweeping the floor at a beauty salon, to help keep the family housed and fed. She was a hard worker, and eventually her employer gave her more responsibilities. Before long, she was promoted to junior hair stylist, and she found that she had a knack for it, especially for the new science of the permanent wave. Her skill in this particular respect did not go unnoticed: She was recruited by the owner of a salon sixty miles away, in Montgomery, Alabama. Soon she was living on her own at an age she guessed was about fourteen, earning enough to take a room in a respectable neighborhood and even to buy a car. She was certain that she'd found her vocation.

But one day a client, struck by her intelligence, suggested she try her hand at nursing. Jackie balked at first, protesting that she lacked the education she'd need to pass her exams. Still, the idea was attractive, and at last she gave in and enrolled in nursing school. For three years she trained and studied. At the end of those three years, however, she failed her licensing exam; she could read but she could not write. Because she had no license, the only physician who would hire her was a practitioner in the Florida backcountry. She gave it a try, but the surroundings were all too depressingly familiar—impoverished patients crowded into tumble-down windowless shacks. Her nursing training had sent her right back into the bleak environment she had worked so hard to escape.

So she gave up nursing, moved to Pensacola, in Florida's panhandle, managed a loan, and bought into a partnership in a local salon. Before long, the business was making a profit, and everything was beginning to seem settled. But if she had found a professional identity, she still lacked a full name. And then one day, running her finger down the pages of the phone book, she discovered a name she really liked: Cochran. She found herself thinking, *It sounds like me. I have my own life now, a new one. What better way to begin than with my own name?* Cochran. *Why the hell not?*

By her twenties, Jackie Cochran had pulled herself up from impoverished mill hand to successful small-business owner. She had accomplished a great deal against long odds, and for many people, all this—owning a business, making a good living—would have been enough. For a while it was enough for Jackie Cochran too.

But that was before she discovered flying.

2

THE GOLDEN AGE

By 1911, eight years after Wilbur and Orville Wright had piloted a heavier-than-air flying machine, a lot had happened in American aviation. The Wright brothers had sold an airplane to the U.S. government, Glenn Curtiss had demonstrated the feasibility of cross-country flying, and the first airmail had been delivered. By 1911, America boasted forty-six licensed pilots. Even the U.S. Army was beginning to see possibilities for airplanes. In April the Army Signal Corps ordered two young lieutenants to Dayton, Ohio, where they learned to fly under Orville Wright. Six weeks later, they were dispatched to College Park, Maryland, to instruct other Army pilots.

By 1911, many Americans saw promise for great things in the future of aviation. Some predicted that this future would belong to everyone. That year, L. Frank Baum, author of the bestselling *Wizard of Oz*, published a novel called *The Flying Girl.* Baum believed that women were destined to make a place for themselves in the sky, and he hoped that young female readers would be inspired to emulate his heroine's daring aerial feats. The book's introduction observed, "the American girl . . . already recognizes her competence to operate successfully any aircraft that a man can manage . . . In America are thousands of girls ambitious to become aviators."

At the time, there was only one licensed woman pilot in America, but Baum's claim about women and flying was already coming true. Some years earlier, a woman named Blanche Stuart Scott had taken lessons from aviation pioneer Glenn Curtiss. She mastered the rudiments, but Curtiss forbade her to fly solo and blocked the throttle. Scott managed to undo Curtiss's fix, and while practicing taxiing, rose from the runway for a few moments and landed again, thus becoming the first American woman to fly an aircraft, albeit unofficially. On another front, Bessica Raiche and her husband had flown brief flights in a biplane they had constructed in their Long Island home, and Bessica had received a medal inscribed to the "First Woman Aviator of America" from the Aeronautical Society.

Soon enough, there were others. In 1912, a writer named Harriet Quimby crossed the English Channel in a fifty-horsepower Blériot monoplane. In 1916, an intensely competitive woman named Ruth Law set a new nonstop distance record, a 590-mile flight from Chicago to Hornell, New York. And Marjorie and Katherine Stinson and their brothers founded a flying school in San Antonio; while Marjorie instructed (she had tutored more than one hundred Canadian pilots before she was twenty-two), Katherine toured Japan and China, performing exhibition flights in a Laird Looper before crowds in the tens of thousands.

When the United States entered World War I in 1917, civilian flying was banned for the duration, but several women found ways to keep involved. Marjorie Stinson worked as an aeronautical draftsman for the Navy. Anita Snook inspected aircraft engines under production. Ruth Law, having offered her services to the Army as a combat pilot and been refused, flew for Liberty Bond drives.

The war greatly accelerated engineering advances in aeronautics. When it was over, planes no longer looked like fragile box kites. They were larger, more reliable, sturdier. They were also cheaper. In 1918, when hundreds of single-engine open-cockpit Curtiss Jennies were declared surplus, former Army pilots, seeking a way to stay airborne, leased them. The pilots flew around the country, appearing at public exhibitions and country fairs, selling airplane rides, and thrilling spectators with loops and rolls. By the 1920s, small-town Americans could stroll down the street from their homes on a fine summer afternoon and see the future unfolding.

Quite a few of the barnstormers were female. One was a black woman named Bessie Coleman. When she was refused a license in the United States, Coleman earned hers in France, and returned to America to perform, the color of her skin suddenly a novelty and a draw. There was also Phoebe Hargrave and her Flying Circus. Phoebe performed double-parachute jumps in which she cut loose the first chute, free-fell for several seconds, then opened a second chute which carried her to earth. On summer nights Katherine Stinson would put lights on the wingtips of her plane and perform aerobatics in the dark. And then there was Mabel Cody, Queen of the Air. Mabel Cody would dangle from the plane's spreader bar with one hand, and transfer from one plane to another.

In the 1920s, thousands of American girls who saw Mabel Cody or Katherine Stinson or Bessie Coleman never looked at the sky the same way again. Many of them would try to imagine flying, and some of them determined that they would. Flying was hardly a typical ambition for a woman, but the social climate seemed to favor women who wanted to break the mold. Americans had become attracted to the idea of the rebellious, nonconformist woman. It was the era of the flapper—the bold, independent woman whose boyishly slim figure bespoke her rejection of traditional femininity—and of the New Woman, who was impressively clever and competent. In the 1920s, women in unprecedented numbers were moving into fields, like law, medicine, business, and the sciences, that had traditionally been reserved for men. And in some ways the aviatrix was the perfect icon of the "new" femininity. She was the flapper and the career woman rolled into one: a wild adventuress and a serious, skilled master of a challenging profession. She was participating in the realization of incredible new possibilities. She was brave and she was enterprising. And she did not let traditional notions of what she was "supposed" to do stop her. In short, she was just about everything Americans wanted their heroes to be.*

*The women understood the excitement and glamour that they aroused, and played to the crowds. Harriet Quimby had fashioned herself a hooded flying suit made of plum-colored silk. A wing walker born with the name Margie Hobbs dubbed herself "Ethel Dare, the Flying Witch." Male pilots also learned to appeal to a public's sense of aviator romance. Roscoe Turner had a waxed mustache and a British flying tunic and pilot's wings of

But the girls who decided to become pilots soon discovered that public response to female fliers was mixed. Flying was not ladylike. It was dirty, greasy, and noisy. It required an understanding of engines and cables and control surfaces, and it was difficult to do in a skirt. Mothers of aspiring aviatrixes would worry that their daughters would never find husbands. It was all very well to be adventurous, but it was no laughing matter for a girl to risk ending up a spinster because she persisted in acting too much like a man. And for all but the wealthy there were financial barriers. Ordinary flying—managing a school or carrying passengers—offered a precarious income at best, and many pilots were forced into stunt flying. Although some stunt flyers made considerable profits (Katherine Stinson's stunts raised a great deal of money for the family's school, and Bessie Coleman expected to retire on her savings and found a school for black pilots), most stunt fliers lived hand to mouth. Maintenance, repairs and fuel were constant worries.*

There was another reason a mother might want her daughter to find another career, and it had little to do with marriage prospects or fiscal solvency. In formal photographic portraits of the period it was easy to spot the aviators: They were the ones wearing bandages or leg casts, or standing on crutches. Injuries were particularly common in stunt flying, especially because by the mid-1920s, the public had become jaded, and expected more and more dangerous stunts. A crowd which had witnessed a plane-to-plane transfer in June would be satisfied only with an automobile-to-plane transfer in July. Many stunt flyers began wing walking or parachuting, although some flatly refused. Matilde Moisant responded to her family's fears for her safety and stopped flying a year after her first flight. Blanche Scott also retired, telling reporters, "Too often, people paid money to see me risk my neck, more as a freak—a woman freak pilot—than as a skilled flyer."

Both had survived crashes. Others were not so lucky. In 1917

---

fourteen-karat gold with diamond studs. For six months he flew with a pet lion cub named Gilmore in the copilot's seat.

*In the 1920s other women, nonpilots, looked to barnstorming as the only way they could become airborne. Gladys Roy would dance the Charleston on the wing. A onetime waitress named Lillian Boyer would ride in the passenger seat of a race car around a track chasing a plane flying overhead. A rope ladder would drop from the plane, and Lillian would climb the ladder into the plane.

Harriet Quimby was thrown from her plane over Boston Harbor, and fell to her death. Others would be killed in stunt accidents—Laura Bromwell in 1921 and Bessie Coleman in 1926.

Meanwhile, male pilots were setting new records in distance, altitude, speed, and endurance. In 1924 Lowell Smith and Erik Nelson flew two Army Air Corps Douglas observation planes around the world. There were flights from New York to California accomplished between sunrise and sunset. And there were new records in altitude and duration, so many that the public was getting as bored with aeronautical records as it was with wing walkers.

Then, in May 1927, a pilot named Charles Augustus Lindbergh flew alone from New York to Paris. His was not the first trans-Atlantic nonstop flight—England's Captain John Alcock and Lieutenant Arthur Whitten Brown had first made the crossing in 1919—but it was the longest nonstop solo flight over any ocean, a feat that many had said was impossible. Lindbergh returned to America a hero. If word got around an airstrip that Lindbergh was dropping in, crowds would race out onto the field, pushing and jostling. And more than one unlucky pilot who made use of the same runway at the same time ended up with a cracked wing, a broken propeller, or even an overturned plane.

The cheers for Lindbergh had barely died in October 1927, when twenty-three-year-old student pilot Ruth Elder and flight instructor George Haldeman took off from Roosevelt Field bound for Europe in a Stinson Detroiter they had christened *American Girl*. Several hours into the trip, their oil pressure dropped to near zero. They came down in open ocean northeast of the Azores, and were rescued by the crew of a Dutch freighter.

Meanwhile, Frances Grayson, a wealthy real estate broker from Long Island, had determined to be the first woman to cross the Atlantic, if not as a pilot then as a passenger. On December 23 her Sikorsky S-36 Amphibian took off from Roosevelt Field with a crew of three, headed north to Newfoundland. It was last sighted off Cape Cod, and radio contact was lost. A prolonged search would find no trace of plane or crew.

Before 1927 was out, nineteen men and women had died attempting Atlantic crossings, and still there were plans for more. In

1928 Mrs. Frederick Guest of London purchased a three-engine Fokker she renamed *Friendship*, and hired a pilot to carry her eastward across the Atlantic. But in deference to her family's worries about her safety, she consented not to make the trip herself but rather to find a suitable surrogate. The committee she appointed to find this surrogate made the offer to a thirty-year-old social worker named Amelia Earhart, who spent weekends flying out of an airport near Boston. The social worker accepted, and in June 1928 she accompanied Wilmur Stultz and Louis Gordon on a transatlantic flight from Trepassey, Newfoundland, to Burry Port, Wales. She was a licensed pilot, but because she had no experience with multi-engine aircraft or instrument flying, and because the flight was mostly through fog, they did not allow her to touch the controls once for the twenty-hour-and-forty-minute crossing. After the flight, few noticed pilots Stultz and Gordon, and fewer would remember their names. But the name of their passenger would take a permanent place in history.

Amelia Earhart had poise and a boyish but patrician beauty, and she looked so much like Lindbergh that she might have been his sister. The newspapers began calling her "Lady Lindy," "The First Lady of the Air," and "America's Sweetheart of the Air." She and the crew of *Friendship* remained in England for two weeks; there were teas, speech-making, and an appearance before Parliament. Upon their return to the States they were offered a tour of thirty-two cities. She was slightly embarrassed by the attention, and of her role on the crossing she remarked that she had been "as useful as a sack of potatoes." When the crowds heard such modesty they just cheered louder. Amelia Earhart had captured the imagination of America. She had also captured its heart.

Throughout the 1920s, American women aviators were experiencing various degrees of success, operating flying schools, flying charter flights, selling and servicing airplanes. But none of this attracted much attention, and the public and male flyers were generally suspicious of the female pilots. The situation made some women all the more determined to prove that the airplane didn't care who was flying it. They knew they could demonstrate their skills dramatically by breaking records recognized by the Fédération Aéronautique Internationale (FAI) or its American representative, the National Aeronautic Association. Viola Gentry, Evelyn "Bobbi" Trout, and Elinor Smith all set women's records in endurance and distance. A still more dramatic demonstration of flying skills would come with the first races.

In August 1929 many women pilots had never met another pilot of their own sex, so there was an electricity in the air on the eighteenth in Santa Monica, California, when twenty women pilots began the first Women's Air Derby, an eight-day race to Cleveland, Ohio. The women flew a series of legs of several hundred miles each, over desert and mountains, navigating only with compasses and roadmaps. Amelia Earhart and Mary von Mach experienced engine trouble, and Florence "Pancho" Barnes crashed on landing in Pecos, Texas. The women arrived at an overnight stop in Douglas, Arizona, to learn that Marvel Crosson, flying above the Arizona desert, had lost engine power and bailed out at 2,000 feet, too near the ground for her parachute to open. Immediately there were calls for an end to the race, one newspaper headline proclaiming "WOMEN HAVE PROVEN THEY CANNOT FLY," but derby manager Frank Copeland flatly refused, as did the race committee. The criticism merely galvanized the remaining contestants, who flew on. On August 26, 1929, Louise Thaden won the race, Gladys O'Donnell ran second, Amelia Earhart ran third, and eleven others finished. There had been a number of minor accidents due to mechanical failures, and of course there was shock at Crosson's death. But there was also a new and fierce camaraderie among the women. They decided to organize.

On the rainy night of November 2, 1929, twenty-six women pilots arrived at Curtiss Airport in Valley Stream, Long Island, in order to form a national organization of women pilots. They met in a room above a hangar where mechanics were running up a Curtiss Challenger engine, and they almost had to shout to hear each other. But it was quite a sight. Twenty-six aviators, all of them women and many of them famous. They served tea and biscuits on a toolbox wagon. They elected Neva Paris temporary chairperson, drew up a statement of purpose which included: "good fellowship, jobs, and a central office on women in aviation." There were a lot of suggestions as to what to call themselves—the *Noisy Birdwomen*, the *Homing Pigeons*, and so on. Then

Amelia Earhart and Jean Hoyt suggested the name simply be taken from the sum total of the members. In the subsequent weeks the remaining applications arrived, and on the day of the deadline the membership committee counted exactly ninety-nine, so they became the *Ninety-Nines*.

3

EPIPHANY

By 1929, Jackie Cochran was co-owner of a thriving beauty salon in Pensacola. Life was comfortable and there were lots of dates with handsome young men from the new Pensacola Naval Flying School. But somehow it wasn't enough. It was too easy. It was as though she could see a whole life stretching out in front of her as though it had already been lived.

So she moved to New York City. She knew that in the beauty business, Manhattan was the place to make a name, not just locally but nationally. And she had ambition in that direction. Her idea was that one day she would launch her own company, selling beauty products coast to coast. "Jacqueline Cochran Cosmetics" had a nice ring to it, but she didn't have the money for that kind of enterprise yet, so she took a job as a hairdresser at Antoine's Salon at Saks Fifth Avenue. For three years, she styled hair and saved money. At the end of those three years she still lacked the necessary start-up capital. So by the spring of 1932 she was considering leaving Antoine's for a job as a traveling sales representative for a cosmetics company like the one she intended to own herself one day. She would learn the business from the ground up. It was a sensible plan in most respects. Except that in 1932, in the richest country in the world, more than five million men were looking for jobs that did not exist. Cases of starvation were being chronicled by all the major newspapers, and families who a few years earlier had been respectable were begging in the streets or ransacking garbage cans. It would become known as the cruelest year of the Depression. It was a bad time to give up a secure job. So Jackie wasn't sure.

But there were many who survived in that year, and some who actually prospered. Even in 1932, the patrons of Antoine's were women of considerable means. Many of these women wintered in Florida, and because Antoine's had a shop in Miami Beach, Jackie Cochran went with them. Sometimes Jackie and the women she worked for would become friends, and she would find herself invited to dinner parties where she was seated next to people she had read about in the society columns back home.

At one such party, on a warm spring night in Miami, she was introduced to a man who seemed a good deal younger than most of the other dinner guest tycoons. He was blond and freckled and unassuming. He told her he was the son of a midwestern Methodist minister. His name was Floyd Odlum. He had a nice smile and a pleasant manner, and he became even more attentive to Jackie when he found out that she was a beautician. He said he enjoyed meeting working women, because their lives were so interesting. He asked her so many questions about her job that she didn't get a chance to ask him what he did. She thought maybe he was a small-town lawyer or business man. She was disappointed to learn that he was married, and a little surprised when she realized she was disappointed. He kept asking her questions about the beauty business, and the next thing she knew she was telling him about her idea of selling cosmetics. He listened attentively, and then he said something both startling and compelling: "If you're going to cover the territory you need to cover in order to make money in this kind of economic climate, you'll need wings. Get your pilot's license."

Jackie just looked at him. She didn't quite know what to say. Was he joking with her? Become a pilot so she could sell lipstick and pancake? It sounded crazy. But it also made a crazy kind of sense. The idea sat there in the back of her mind, an intriguing possibility. By the end of the evening, she had learned that this pleasant man was someone whose business advice was to be taken seriously. Floyd Odlum was a Wall Street lawyer and investment genius; barely forty, he was already one of the wealthiest men in the country. He had hardly felt the effects of the bank closings. Any of the other tycoons in the room that night could have told her: If you want to get ahead in the world, and Floyd Odlum says you can do it by learning to fly, then sign yourself up for lessons tomorrow. But Jackie didn't ask the other tycoons for their advice. She just let the idea sit there in the back of her mind, and every now and then she turned it over.

Jackie and Floyd kept in touch. He lived in New York too, where he kept offices, and occasionally they enjoyed each other's company over drinks or dinner. One day he asked her if she had thought any more about that pilot's license. She hesitated, and he could see that she was on the fence about it, so he smiled and said, "Okay, Jackie, I'll bet you the cost of the lessons that you can't get that license in six weeks." That did it, as he knew it would. Jackie accepted Floyd's wager, but she upped the ante. She was going to become a pilot in only three weeks, she informed him, because she had other things to do with her vacation time.

Her instructor, Husky Lewelleyn of the Roosevelt Flying School on Long Island, held a low opinion of lady pilots, and he just laughed when Jackie told him she would require his services for only twenty-one days. But he stopped laughing about five minutes after she took the controls of the sixty-horsepower Fleet trainer. Lewelleyn knew right away that Jackie Cochran was one of those rare people that he referred to, with awe, as born pilots. After that first flight, it didn't surprise him one bit that Jackie earned her license in exactly three weeks, just as she said she would.

Jackie started flying as nothing more than a means to an end. But after only three weeks, it had become an end in itself. Her first time in the air had been a kind of epiphany: flying, she found, challenged her awareness and her reflexes in an exhilarating way. When her friends asked her what it was like, she could only describe it as a "stupendous sensation." She was hooked, but she realized that she still had plenty to learn. So in 1933 she quit her job and moved to San Diego, where she enrolled in the Ryan School of Flying. By the end of the year, she had earned a commercial license and scraped up the twelve hundred dollars necessary to buy an airplane of her own, an old Travel Air. Aviator friends from Pensacola gave her an informal but intensive flight-training course.

Jackie had a lot of confidence in her abilities as a pilot, but as far as the rest of the aviation world was concerned, she was still an unknown. The way to make a name for herself was in racing. In 1934 she entered the MacRobertson Race from London to Melbourne, Australia. She was in first place when her plane, a GeeBee Racer R-1, broke down in Bucharest. She did not finish. But a few months later she was ready for the 1935 Los Angeles-to-Cleveland Bendix Race. By then she had met the only other woman contestant—Amelia Earhart.

By 1935, Amelia Earhart had flown solo across the Atlantic, and had become the first pilot to make a solo trip from Hawaii to the American mainland, the first to fly nonstop from California to Mexico City and the first to fly nonstop from Mexico City to New York. She was considered one of the best pilots in the world. In the 1930s, while women pilots competed against each other in all-female air races, few women participated in the most prestigious races like the MacRobertson and the Bendix. In fact, with the exception of Amelia Earhart, those women who tried to sign up for these races often met with resistance. Such was the case when Jackie decided to enter the 1935 Bendix. In 1933, a woman contestant in the Bendix had been killed, and the race's officials declared that they would not be responsible for more female lives. But when Amelia Earhart heard that Jackie was having trouble gaining a spot in the Bendix, she entered the race herself. Thereafter, the officials gave Jackie no more trouble.

Amelia Earhart finished fifth in the Bendix that year. Jackie's Northrop Gamma overheated, and she was forced to quit in Arizona. But it was Jackie who, three years later, would place third in the 1937 Bendix. In 1938 she would enter the Bendix again—this time in an experimental long-range Seversky AP-7 pursuit plane—and she would take first place. In 1939 she set a new women's altitude record, and in 1940 two absolute speed records. By this time Amelia Earhart, whom Jackie had come to regard as one of her closest friends, was . . . missing.

In the summer of 1937 the disappearance of Earhart's specially equipped Lockheed Electra near Howland Island in the Pacific had galvanized the United States into the most exhaustive search in its history. While the American public began to tell itself stories about how Earhart had vanished on purpose, and while these stories were evolving into myths that would survive throughout the century, Jackie Cochran knew, with infinite sadness, that her friend was dead.

Floyd Odlum was a wealthy man, but he seldom acted like one. In his leisure time, he did not play tennis or drink martinis, but preferred to relax after a hectic day at the office by sitting alone in a little room in his vast house, where he would model figures out of clay. Floyd once said that he'd made so many of these figures over the years that he could have filled a whole room with them. But to him, such a display would have been vain, and he destroyed his creations as soon as they were completed.

By 1935, Floyd Odlum was going through a divorce, made public and all the more painful by the scandal-sheets, but he had also found a new hobby that served as quite an effective distraction from the prying eyes and ears of the public. When Jackie was preparing for a big race, Floyd would plunge right into the preparations too. He'd peer through his horn-rimmed glasses at maps and charts, plotting routes, and studying fuel consumption rates and aerodynamics and engine capacities, suggesting an altitude change here and a burst of speed there. The prospect of piloting a plane himself in a difficult race against stiff competition did not appeal to him, but he found the strategizing utterly engrossing. Floyd Odlum did not have a pilot's athletic prowess, but he did have a fine pilot's mind.

Their personalities could not have been more different, but underneath Jackie's brash demeanor and Floyd's gentle one lurked the same kind of fierce intelligence, the same driving work ethic, and the same ambition to be the best at whatever they did. And in 1936, in a quiet ceremony in Kingman, Arizona, they were married. When an orphan marries a millionaire, there is always talk of a certain kind, and at first there was speculation about why a vivacious young gal like Jackie Cochran might have married quiet Floyd Odlum. But after a while the talk died down, because when it came to Jackie and Floyd, there was just nothing to talk about except the fact that they were plainly very much in love.

They moved into a spacious apartment in Manhattan that was near the building that housed a fledgling enterprise called Jacqueline Cochran Cosmetics. Out in the California desert they began building another home, a vast estate that would be known as the Cochran-Odlum ranch. When they were married Floyd gave Jackie a sealed envelope, saying, "Your parents' names are recorded here—if you ever want to know." He knew she'd been raised by a foster family and knew nothing of her birth parents. Because she had once said to him that it might be nice to know, Floyd had hired a private detective. Floyd told her he had not read the papers inside the envelope, and it didn't matter to him whether Jackie ever did. What mattered was that the knowledge was there for her, if she wanted it.

Jackie did not open the envelope that day, and she did not open it on any of the subsequent days. But she kept it, almost like a talisman, not because of the past it could reveal but simply because Floyd had given it to her. With Floyd in her life, Jackie found that the past she'd so often wondered about just didn't seem important. The envelope remained unopened.

4

THE ARMY WAY

The two second lieutenants the Army had sent to the Wright Company Flying School in 1911 were Thomas D. Milling and Henry H. Arnold. When Arnold completed the six-week course, he was certified as U.S. Army Aviator Number Two, and posted as an instructor at the Army Signal Corps' flight school in College Park, Maryland. The Army regarded its foray into aviation as a kind of experiment, and for some time the flight school had more airplanes than it had pilots. So the instructions Lieutenant Arnold received about precisely what it was that Army aviators were supposed to do were somewhat vague. One conceivable use for them was reconnaissance. The idea was something like this: During a battle, a pilot could fly up over the action and make note of any tactical information not obvious from the ground. He could jot down his observations on brightly colored pieces of paper and drop the weighted paper to the ground, where a Cavalryman, galloping hell for leather, would pick it up and deliver it to the command post. To young officers with high aspirations, a stint in the Air Service hardly looked like the way to begin a stellar military career.

But World War I demonstrated that airplanes could in fact be useful in combat. The U.S. Army discovered aerial observation, aerial combat, air support of infantry, and strategic bombing. Some saw aircraft as humane, expecting that they might shorten a war. Even so, many in the Army remained unimpressed. Prevailing sentiment was that flying was an interesting hobby—like fencing, badminton or collecting first editions—but that it held little promise for waging war.

World War I and its confused aftermath contributed to widespread isolationist sentiment in America. The United States Senate refused to endorse membership in the postwar League of Nations, fearing that it would entangle the U.S. in infighting among foreign nations. By the mid-1930s, as parts of Europe fell under the shadow of dictatorships, Congress passed a series of neutrality acts. The Army's plans for national defense centered only on a small regular force guarding borders. A country with friendly nations on its northern and southern boundaries, and vast oceans to the east and west, seemed to have little need for an extensive Air Corps. Funding for the Air Corps, consequently, was erratic during the Depression. In 1936 the inventory fell to 855 aircraft and 1,500 officers. Between 1919 and 1939, the largest number of pilot training graduates in one year was 246. Even in 1938, when developments in Europe prompted more outlays for the Army, ground forces were given priority.

On the eve of World War II, the Army Air Corps was like an unwanted stepchild of the Army, and most military officials never thought it would be much more. For a long time, no one even knew what to call it. Between 1911 and 1918, it evolved from the Air Service to the Air Division to the Aeronautical Division, to the Airplane Division, to the Air Service Division, to the Aviation Section, to the Air Service of the National Army (ASNA). If its officers weren't being given the new planes or the manpower they were clamoring for, at least they were kept busy putting in orders for new letterhead. But moving up to service status did mark a token acknowledgment of the airplane's potential importance. Previously it had been nothing more than a branch of a branch—sort of a twig, a subsidiary of the Signal Corps. As the ASNA, the Air Service actually became a full-fledged branch of the Army. But that wasn't the end of the name changes. In 1927 the Air Service of the

National Army became the Army Air Corps. The Army Air Corps was still the Army Air Corps over a decade later in 1938, when Henry H. "Hap" Arnold became its chief.

In 1938 Hap Arnold was fifty-three, and he looked like the most contented of small-town doctors. His energy was astonishing, and as was typical of Air Corps officers, he was promoted quickly. During World War I he had become assistant director of the Office of Military Aeronautics, and after several field commands in the 1920s, he organized and led a group of B-10 bombers from Washington, D.C., to Alaska and back. By 1936 Arnold was assistant chief of the Air Corps. With the death of Major General Oscar Westover in 1938, Arnold became chief of the Air Corps.*

Arnold believed that war was inevitable, and he believed that aircraft would prove invaluable. He knew that the United States was poorly equipped to wage an airwar, especially in relation to the air forces of the major European powers. The Army Air Corps maintained eight-hundred usable aircraft. Meanwhile, Britain had two thousand, and France seventeen hundred. Germany's Luftwaffe had thirty-six hundred.

At that time, Hitler had assimilated Austria into what he was calling the Third Reich. By fall, his forces had taken Czechoslovakia. In a secret meeting on November 14, 1938, six weeks after the Czech surrender, Roosevelt outlined a program for aerial expansion, describing an Air Corps of 20,000 planes, and an industry which could manufacture 24,000 more each year. On January 12, 1939, he asked Congress to make $300 million available to the Air Corps. Most of it would bring the Air Corps's inventory to 5,500, and the remainder would pay for men to operate and maintain equipment, new training facilities, and new operating bases.

But there was still the matter of pilots. Arnold believed that, in the event that the U.S. went to war, the Air Corps would need twelve hundred new pilots annually. The difficulty was that, in an isolationist atmosphere, such a massive increase of military pilot training would attract unfavorable attention. Arnold's task was to

*Arnold's career was not without controversy. Like many young Air Corps officers, he had supported Brigadier General William "Billy" Mitchell during his court-martial for insubordination in late 1925. Arnold shared Mitchell's view that the Air Force should be separated from the Army; in fact, Arnold was himself nearly court-martialed in 1926 for using Army paper and equipment to disseminate pro-airpower views.

prepare for a war that, as far as the American public was concerned, was not going to happen. So he arranged for civilian flying schools, beginning in mid-1939, to provide primary flight training to any college student paying a forty-dollar fee. It was called the Civilian Pilot Training Program (CPTP). The Air Corps supplied the aircraft and supervisors; the instructors were civilians. Ostensibly, the program would help young Americans take part in the new Air Age. But when Arnold needed his pilots, the CPTP would be able to provide them.

Because it was a civilian program, it was open to all, regardless of race or sex. And so it happened that a substantial number of CPTP students—roughly three thousand—were young women. For these women, the CPTP meant inexpensive flying lessons, and a wonderful opportunity. But to the Air Corps, the women students were nothing more than a kind of expensive camouflage—evidence that the CPTP did not exist solely to train pilots for war. Because while these women students received the same training that their male counterparts did, and while women, in small numbers, did have a part in civilian aviation, it was plain to many that putting a woman in the cockpit of a military plane was simply unthinkable.

5

NANCY

In May 1940 Colonel Robert Olds, chief of the Army Air Corps Ferrying Command, received a letter from a young pilot named Nancy Harkness Love. In it, she suggested that the Air Corps might be interested to know that if there were a war, women aviators could be of considerable use as ferrying pilots, delivering airplanes from factories to air bases and from one base to another. "I've been able to find forty-nine I can rate as excellent material," Love wrote, ". . . there are probably at least fifteen more. I really think this list is up to handling pretty complicated stuff." This idea was unusual, to say the least. But its source was impeccable. As pilots went, Nancy Love was pretty "excellent material" herself.

Nancy Love was from a Philadelphia family the newspaper society pages called prominent, and she had attended the right schools—Milton Academy in Massachusetts, then Vassar in upstate New York. Nancy was articulate and witty, and particularly striking because her hair had gone silver while she was in her twenties. Her voice was a whiskey contralto she could use to great effect. But some said she didn't always seem to know how to act properly. At Vassar she had made pocket money by giving rides in a plane she rented at the Poughkeepsie Airport, and one fine weekend she took a few friends up for a proficiency run, and buzzed the campus at treetop level, low enough that somebody on the ground read the tail number and reported it to the airport. College officials suspended her from school for two weeks, and from flying for the remainder of the semester.

For the rest of that year there were jokes about it, and her classmates started calling her the "Flying Freshman." Nancy had had enough. She resigned from Vassar at the end of her sophomore year. It was 1934. There were few jobs for experienced pilots, and Nancy had only three hundred hours of flying time. But she used what she had: a will, with which she found work at East Boston Airport selling airplanes at commission; and her family, who could offer her a too-short list of acquaintances who had not been ruined by the bank closings. On that list was a millionaire who seemed mildly interested in several planes, but Nancy couldn't get him to bite. In fact, he was not as interested in the planes as he was in the woman selling them. Specifically, he was in the market for a "suitable wife" for his eldest son. Nancy knew nothing about all this—and Joseph Kennedy Sr. didn't know that as far as marriage was concerned, Nancy had plans of her own. . . .

His name was Robert M. Love. He was a pilot and an officer in the Air Corps Reserves. Nancy had met him at East Boston Airport, where he owned a fledgling airline called Inter City Aviation. When they were married in 1936, Nancy became a kind of local celebrity. Boston papers were entranced with the mere idea of Nancy and Bob. BEAUTIFUL AVIATRIX WEDS DASHING AIR CORPS OFFICER . . . THE ROMANCE OF THE GLAMOROUS YOUNG SOCIETY COUPLE MEETS THE ROMANCE OF THE SKY . . . Nancy was a bit embarrassed by the publicity, but she knew how to be gracious and polite, and she learned how to become detached—some said aloof. It sometimes seemed that she was something of an enigma, that beneath the gracious exterior, it was impossible to tell who the real Nancy Love was. In fact though, there was nothing mysterious at all about the real Nancy Love.

The real Nancy Love was a pilot.

By May 1940, Hitler's armies had overrun Poland and Scandinavia, and they were beginning campaigns in France and the Low Countries of western Europe. Great Britain was facing the prospect of an aerial onslaught, and U.S. involvement in the war seemed more and more likely. It was then that it occurred to Nancy Love that she—and perhaps a dozen pilots like her—could ferry aircraft for the Air Corps. Talented pilots looking for work, an Air Corps looking for talented pilots—it had all the makings of the most symbiotic of relationships.

In May 1940 Colonel Olds read Nancy's letter, and was intrigued by the idea of women ferry pilots. In the abstract, it made sense. In the abstract, a pilot was a pilot, and a pilot who had a lot of hours logged on a lot of different kinds of planes was a pilot the Air Corps could use. But as soon as he took this particular idea out of the realm of the abstract, he was back in the realm of . . . women in the cockpits of military planes. Olds decided the idea was a little too radical for the Air Corps, and he declined her offer. But he filed her letter away for future reference all the same. Nancy Love took Olds's response philosophically. If the Air Corps wanted to wait and see, she could do the same.

But Nancy Love wasn't the only person imagining women in the cockpits of military planes. Nor was she the first to propose it. In September 1939, the day after Warsaw fell to Hitler, Jackie Cochran had written a letter to Eleanor Roosevelt, whom she knew slightly, advancing a similar plan. Jackie's idea, however, was a little more ambitious than Love's. She suggested that women pilots could do a great deal for the Air Corps—administrative flying, testing, ferrying, everything except combat missions. She proposed that the Air Corps make plans and provisions to hire at least 650 female pilots. And she noted that the Air Corps would need to move quickly if it wanted the program up and running by the time there was a war. Eleanor Roosevelt thanked Cochran for her proposal, but neither she nor Cochran ventured to take it, as an official proposal, to anyone in the Air Corps. At least not yet. Like Nancy Love, Jackie Cochran was willing to wait.

In the spring of 1941, at an aviation awards ceremony in Washington, Jackie saw her chance to make a case for women military pilots again. By the spring of 1941, war was imminent, and Jackie knew the Air Corps still faced a shortage of skilled pilots. At the awards ceremony, she ran into Hap Arnold, with whom she had become acquainted when both served on the nominating committee for an aviation award called the Collier Trophy in 1937. Jackie knew that the months ahead would be difficult for Arnold, and she volunteered to do anything she could to help. Arnold eyed her thoughtfully and decided that perhaps there was something Jackie could do for the Air Corps. Then he drew her into a corner and outlined his plan.

In March 1941 Roosevelt persuaded Congress to pass the Lend-Lease Act, making weapons available to Great Britain and other nations, requiring only that remaining, intact weapons be returned after the war. In 1941 one of Hap Arnold's biggest headaches was delivering American planes across the Atlantic to the British. When pilots and aviators talked about this flight they called it "jumping the pond." The phrase had a casual, jaunty air, as though it described a pleasant and mildly athletic activity, like tossing a baseball around on a summer afternoon. The reality was otherwise. From the easternmost point of the coast of Labrador to the west coast of Ireland was roughly eighteen hundred miles of open ocean—unforgiving of engine failures, pilot and navigator errors. Few American pilots seemed willing to make the crossing. And here was Jackie Cochran, insisting that she would do just about anything to help out. Hap Arnold told her that if she ferried a plane to England, other pilots might be persuaded to follow.

Jackie Cochran had not forgotten her proposal, but decided not to mention her idea to Arnold, at least, not just then. But it occurred to her that if her crossing might give male pilots the courage to jump the pond, it might also give the Army Air Corps the courage to recruit women.

On June 17, 1941, Jackie and a crew of three flew a Lockheed Hudson bomber from Montreal to Prestwick, Scotland. Twelve hours after their departure they landed at Prestwick safely, and Jackie emerged from the plane with bags full of fresh fruit which she offered to the men at the air base. The British, who had been rationing food for months, seemed almost as grateful for the fruit as they were for the bomber. Over the next few weeks, Jackie had several meetings with British commander Pauline Gower, who, a year earlier, had been authorized to form a women's division of Britain's Air Transport Auxiliary (ATA). Throughout the winter of 1940 Gower's eight recruits had ferried single-engine Tiger Moth trainers to airfields in Scotland and the north of England. Gower briefed Jackie on her work with women pilots, and showed her a small program that was successful.

When Jackie returned home on July 1, she found that her flight had drawn the attention of at least one American policy-maker. On July 2, President and Mrs. Roosevelt, as a kind of thank-you, invited her to lunch at their estate at Hyde Park. Jackie decided that perhaps the time had come. She told the president she had an idea. She wondered if he knew that there were some fifteen hundred to two thousand women pilots in America who could be organized to perform all kinds of domestic flying jobs so that more of America's male pilots could be released for overseas combat duty. Roosevelt said he thought this might be worth looking into. He gave Jackie a note of introduction to the assistant secretary of war for air, Robert A. Lovett. The note informed the assistant secretary that Roosevelt wanted Jacqueline Cochran to research and write up a plan for establishing an organization of women pilots to fly for the Army Air Corps.*

Jackie, with her endorsement from the president, was ready to put her plan in motion. Lovett named her a consultant, offering her office space in the Ferrying Command. Hap Arnold introduced her to her new boss, Colonel Olds, who had filed away Nancy Love's letter the year before. Her first task was to determine exactly how many current women pilots might be qualified for Air Corps service, which meant sorting through the records of the three hundred thousand Americans who had ever earned a pilot's license. She'd have to figure out who among them were women, who was alive, and who had enough flying hours to do the military any good. This was a enormous task, and Jackie had no intention of shutting herself away among the file cabinets for weeks on end. So she brought in seven employees from Jacqueline Cochran Cosmetics to do the legwork.

In the records of the Civil Aeronautics Administration (CAA) they found the names of some three thousand American women pilots, but learned that only about one hundred had logged the three hundred hours of flying time required by the Ferrying Command. Because the CAA records did not describe types of aircraft flown or cross-country experience, Jackie and her staff mailed a

*In June 1941 the Army Air Corps was renamed the Army Air Force, a designation it would retain until 1947, when it was made a separate branch of the services. For the first months of the war, official and unofficial communications used the names Army Air Corps and Army Air Force interchangeably.

confidential questionnaire to the hundred women. The response was swift and enthusiastic. And on July 21 Cochran submitted to Colonel Olds a formal proposal titled "Organization of a Women Pilots Division of the Air Corps Ferrying Command." A small group of women, culled from those who answered her questionnaire, could ferry trainer aircraft from factories to airbases. If they proved capable, their number might be increased through the establishment of a pilot training program. Cochran estimated that two thousand women would be prepared to join such a program. On July 30 the proposal was submitted to Hap Arnold.

Meanwhile, Arnold's staff had been performing similar research, and had discovered that there were some seventy thousand male civilian pilots in America. Even accounting for those who would be too old for military service, and others not physically fit for it, this number was significant enough to make Jackie's few thousand women seem unnecessary. So Arnold held that the Army Air Force had no need for female pilots. In the summer of 1941, in fact, the AAF had more pilots than planes. Further, women would require difficult and perhaps costly adjustments in accommodations and training.

Cochran was disappointed with Arnold's decision, but thought that perhaps the general could be made to see things differently if they could talk in person. Arnold agreed to see her.

Hap Arnold got quite a kick out of Jackie Cochran. Like him, she was no-nonsense, earthy, and, when it suited her needs, unconventional. Jackie Cochran considered herself a devout Catholic, yet she possessed an extensive and varied repertoire of curses, which she used as circumstances required. For his part, Hap Arnold had something of a temper, and when he got worked up he had a habit of pounding his fist on his desk and yelling "Goddamn it!" The first time he went into this routine in front of Jackie, she looked him calmly in the eye and said, "Shut up. You'll go to hell if you keep talking like that." If her response was a bit hypocritical, she didn't seem to notice. And Hap Arnold not only shut up, he smiled. As to her proposal for women flying for the AAF, he admired her tenacity, and told Jackie her idea had definite possibilities, but that the time was not right. He made a counter-offer. He suggested that she round up a group of American women to ferry aircraft in Britain for Pauline Gower's division of the Air Transport

Auxiliary. The British need for pilots was far greater. Jackie would have a chance to study the British organization, by all accounts successful in solving problems of women at all-male facilities. In the event that the U.S. Army Air Force did need to call on the resources of female pilots, she would know how to get the whole show up and running. To Jackie, this was only half a loaf, but she took it. She spent much of the rest of that summer and fall recruiting twenty-four experienced women pilots and, with them, made plans to depart for England.

At 7:55 on Sunday morning, December 7, 1941, the first wave of carrier-based Japanese aircraft attacked the Pearl Harbor Naval base and other military installations on the Hawaiian island of Oahu. Word of the raid had already reached Washington when Japanese emissaries severed diplomatic relations. On December 8, the United States declared war on Japan. Three days later Germany and Italy entered the conflict as Japanese allies.

Suddenly, America was at war, and on two fronts. Air Force estimates of pilots and aircraft increased dramatically. And Hap Arnold told Jackie, who was finalizing her plans to go to England, that her idea was still premature at this stage, but it was looking less so. They made a deal: If and when the Air Force decided it was ready for women in its cockpits, Arnold would call her back to the States. At least tacitly, it was understood that if and when the Air Force began a Women Pilots Division, Jackie Cochran would be the woman in charge.

6

THE EXPERIMENT

In June 1942, the organization known as the Air Corps Ferrying Command was redesignated the Air Transport Command (ATC), and placed under the jurisdiction of Brigadier General Harold L. George.* Initially George had recruited pilots from the aircraft services, but when he had run through their ranks, he was still understaffed. Then he turned to private aircraft owners, barnstormers, crop dusters—anyone who could fly a plane and be retrained in multiengines. In the summer of 1942 he had roughly eleven thousand pilots, and it was clear that soon he would need a great many more. And so on September 5, 1942, the ATC sent a telegram to eighty-three American women:

> AFATC S938 PERIOD FERRYING DIVISION AIR TRANSPORT COMMAND IS ESTABLISHING GROUP OF WOMEN PILOTS FOR DOMESTIC FERRYING STOP NECESSARY QUALIFICATIONS ARE HIGH SCHOOL

*With the reorganization of the Army Air Forces begun in March 1942 the Air Transport Command was one of six service commands created to support operational elements of the Army Air Force. Like the other commands, the ATC operated under the jurisdiction of Commanding General Arnold and his Advisory Council. The Ferrying Division of the ATC was charged with ferrying aircraft within the continental United States.

EDUCATION AGE BETWEEN TWENTY ONE AND THIRTY FIVE COMMERCIAL LICENSE FIVE HUNDRED HOURS COMMANDING OFFICER SECOND FERRYING GROUP FERRYING DIVISION AIR TRANSPORT COMMAND NEW CASTLE COUNTY AIRPORT WILMINGTON DELAWARE IF YOU ARE IMMEDIATELY AVAILABLE AND CAN REPORT AT ONCE AT WILMINGTON AT YOUR OWN EXPENSE FOR INTERVIEW AND FLIGHT CHECK STOP BRING TWO LETTERS OF RECOMMENDATION PROOF OF EDUCATION AND FLYING TIME STOP BAKER END GEORGE ARNOLD COMMANDING GENERAL ARMY AIR FORCE WASHINGTON

The ATC had decided to establish a squadron of female ferry pilots. It would be called the Women's Auxiliary Ferrying Division (WAFS), and it would serve as a small, controlled experiment that would allow the Ferrying Division to test the idea of recruiting women to fly Air Force planes.

Male candidates for the Ferrying Division were required to have three years of high school and 200 hours flight time. For female candidates the requirements were significantly more stringent. They had to be high school graduates, with a CAA 200-horsepower rating and 500 hours flight time, at least fifty of which were to be in the previous year. But the newly appointed squadron leader knew there would be more to this than flying airplanes. Her women would be scrutinized by the AAF, the press and the public alike. She wanted their personal conduct to be above reproach. So she raised the bar a little further, asking for two letters of recommendation. Finally, she proposed that the women's salaries be set at $250 per month, $50 less than the men's. Her stated reasoning was that the women would be flying light single-engine trainers, but it had also occurred to her that this would ensure that no male pilot would be unduly threatened.

The hope was that the eighty-three telegrams would yield a squadron of about twenty-five women. The idea had been Nancy Love's, so it was only fitting that Nancy would be the squadron leader. All Ferrying Division pilots were assigned to one of six ferrying groups located at air bases across the country. The WAFS would be part of the Second Ferrying Group at New Castle Air Base in Delaware. By September 6 Nancy was reporting there every day, waiting to see what results the telegrams would bring.

The eighty-three women who received these telegrams were, to say the least, surprised. This was the first any had heard of such a plan. There had been no talk of women pilots in the Air Force in the papers or even on the air fields. And suddenly the Ferrying Division wanted women pilots, and wanted them immediately. Some of the eighty-three had other commitments, and some of them simply did not think it prudent to jump into this on such short notice. But a number decided that, bolt from the blue or not, it was a chance they couldn't pass up. These women made themselves immediately available, and they began to make their way to New Castle.

In September 1942 Betty Gillies of Syosset, Long Island, was thirty-two years old and exactly five feet one-and-one-half inches tall. She was from a family of some means, and she had her own twin-engine amphibian. When she was flying she wore sunglasses with tortoise-shell frames that accentuated her high forehead. Especially when she wore those sunglasses, Betty had the delicate and worldly-wise look of a graduate student at the Sorbonne. Betty and her husband, Bud, had three children, and they lived in a village of great rambling houses covered with weathered gray shingles and fronted with wide wraparound porches. On Sunday afternoons in September, the privileged inhabitants of these homes would stand on those porches sipping the last gin and tonic of the season and watch the fading autumn light play across shaded lawns. The talk would be of getting the boat out of the water.

From their home in Syosset Betty and Bud could drive north two or three miles to the coast and find themselves overlooking Oyster Bay or Cold Spring Harbor. Or, they could drive south the same distance and find themselves at the Grumman Aircraft plant in Bethpage. Most days the Gillies drove south. Bud was one of Grumman's vice presidents, and Betty was one of their utility pilots.

In 1941 their youngest became ill. She was diagnosed with leukemia, and there was little that the doctors could do. She died late in the year, a little before Christmas. The holidays were difficult. Betty just felt numb. Friends began to wonder what to do for her—months passed and she didn't seem to be getting over it. Then in September she received a phone call from Nancy Love. Betty knew Nancy slightly—they had met through friends at the Aviation

Country Club on Long Island. Nancy said "Look, we could really use you down here." Betty hesitated. There were the children, and the work at Grumman. . . . Nancy pled with her, saying, "Stay on just until we get it started—say, ninety days." Betty agreed to think it over. Later that day she mentioned the invitation to some of her friends, and they encouraged her to accept. It could be a kind of therapy, a way to recover from the loss. Bud was all for it.

And so on a September morning on Long Island Sound, Betty Gillies's twin-engine amphibian roared into life, slowly moved out of its slip, and skidded along the surface, gradually gaining speed. Then it rose from the surface of the water, climbed to 2,000 feet, turned, and headed south by southwest.

It was a short flight, and as she began the downwind leg of the landing pattern, Betty could see New Castle Army Air Force Base stretching from horizon to horizon. The base had been built to accommodate the material needs of over ten thousand officers, enlisted men, and civilians. There were rows and rows of barracks, supply depots, mess halls, and administrative offices.

Less than an hour later Betty was sitting in Nancy's office. The room was utilitarian, and it seemed to suit her. As they began to talk Betty remembered that Nancy could be disarmingly candid. "This might not be easy," she said. "Already there are problems. Obviously the women can't use men's barracks—for one thing, the plumbing isn't right. Women's barracks are promised, but because we're civil service, they aren't a priority." She sighed. "For the time being, we'll have to stay in hotels. There are other problems. The finance office has yet to put us on payroll—they're saying it might be a while. And we'll have to pay for our own uniforms." Betty said, "There's a war on, Nance. I'm sure the girls will understand."

Nancy smiled. "You're right, but to be honest, what most concerns me is whether enough women will even sign on. We need only twenty-five. But so far, you and I are it."

Doubts or no, the public announcement about the formation of the Women's Auxiliary Ferrying Squadron occurred three days later, on September 10. That morning Nancy was at the Washington office of Secretary of War Henry L. Stimson. Secretary Stimson hadn't been able to announce much good news lately. The Allies were losing ground in Europe, Africa, and the Pacific. And so this bit of ceremony put him in an unusually good mood. For the small group of admiring reporters, he read parts of Nancy's résumé, and then, as though he were introducing the prom queen, said, "Come, Mrs. Love, will you let the ladies and gentlemen have a look at you?" Mrs. Nancy Love, twenty-eight years old, a Dorothy Lamour look-alike with silver hair pulled tight under a pillbox hat, wearing a tailored suit, stood, smiled graciously, and gave the slightest of bows. The cameras clicked and the flashbulbs popped.

Back at New Castle, others were beginning to arrive. The first was a woman named Cornelia Fort, a flight instructor from Nashville. There was also a tall, plain-spoken woman named Adela Scharr, another flight instructor, from St. Louis. There was a smallish woman named "Pat" Rhonie from New York. Soon there were eight, and then ten. Each was interviewed by an examination board, whose members included the commanding officer of the Second Ferrying Group Colonel Robert Baker, several flight instructors, and Nancy Love. They would flip through her flight log, ask her a few questions about her work in various aircraft and her piloting experience. Then there was a check flight with an instructor. For these ten women, all of this was a formality. Their qualifications were well above the Ferrying Division's minimum standards, and each passed easily.

At first, life at New Castle was a little uncomfortable. At their own expense the ten women stayed at a rooming house in Wilmington called Kent Manor. They stayed there because Nancy was staying there, and simply because nobody had a better idea. At night they ate on base in the new officers' mess. The WAFS had their own table, and there were curious glances from passing lieutenants, sometimes a joke from the back of the room. But when the WAFS began talking, they almost forgot where they were. Most were strangers to one another, and some of them were, at first, a little shy. But they did what strangers do when they are brought together at the dinner table—they asked each other polite questions about their lives, and the questions led to stories, and the stories were mostly about flying, and flying was a subject about which each could never say enough.

As Cornelia spoke, it became obvious that she was quite unlike most members of the Nashville debutante circuit. She loved music and books. She had attended Sarah Lawrence, where she had made friends who helped her find an inner moral fire. Cornelia had written editorials in the college paper advocating entry into the war on ethical grounds. Later, she worked as a flying instructor in Hawaii. On that terrible Sunday morning nine months earlier, Cornelia had actually been flying near Pearl Harbor when the attack began.

Sitting to the left of Cornelia was a tall woman with dark hair and eyes. She was thirty-two, and her name was Adela Scharr, but she would ask people to call her "Del." Her husband, Harold, was a chief petty officer in the Naval Reserve. Del had been a weekend flight instructor and a schoolteacher in St. Louis, and she had a midwestern schoolteacher's belief in honesty, hard work, and thrift. When she was a child her father made ninety dollars a month walking a beat as a police officer, and her mother baked her own bread. They raised chickens and vegetables in the backyard, and had installed their house's steam-heating system themselves.

At the far end of the table was a Civil Air Patrol pilot named Helen Mary Clark. Her husband and her two adolescent sons called her "Angel," and anyone could see why. She had a regal bearing, classic features, and she spoke softly. Seated next to her was a woman who had been an actual barnstormer. Her name was Teresa James, and like most of the others she had gained most of her hours instructing. Now Teresa was telling shaggy dog stories, and she kept at it until even somber Del was laughing.

The large room emptied of officers, but the WAFS, at their table in the corner, just kept talking. The enlisted man on KP duty began to mop the floor, and against the background clatter of somebody washing dishes, the WAFS talked and laughed and talked some more. Sitting at a table over half-empty cups of bitter Army Air Force coffee, the women were feeling as though they'd found long-lost sisters.

The next morning, the WAFS had visitors—reporters and photographers. Seeing the cameras, the women stopped and smiled. And the newspaper people said *Don't stop, go about your business, do whatever it is you do . . . just, you know, act normal.* They wanted pictures of the women doing anything and everything—in flightsuits walking toward the planes, climbing into cockpits, climbing out of cockpits, studying maps in the ready room, sitting on bunks in their quarters. It wasn't long before the Pathé newsreel people showed up in New Castle too. They asked Nancy to pretend that she was greeting her WAFS in front of her office. As they had been there for two weeks, this seemed a little silly, but they played along anyway. Betty Gillies, Teresa James, Cornelia Fort, and Helen Mary Clark put on their civvies and got their suitcases out from the closet and carried them, empty, to the headquarters. Wearing suits and high-heeled shoes, they stood on the ground at the base of the stairs as though they had been dropped off by the same bus. They looked for all the world like four unusually well-dressed and unusually cheerful little girls left on the orphanage steps on the day before Christmas. And Nancy, in her WAFS uniform, stood on the second stair. The director said, "Action," and the camera rolled, and they talked a little and then Nancy stood back against the railing and they all carried their empty suitcases past her, up the stairs. When the newsreel appeared in theaters on the following week it was entitled "GIRLS ORGANIZE ARMY FERRYING UNIT," and the announcer told the story over the little orphanage steps pantomime, and then said, "What will they think of next?" It was the comment of the man at the breakfast table reading an item from the morning newspaper aloud to his wife and then chuckling a little and shaking his head as he turned to the baseball scores.

Soon the WAFS were written up in *Look* and in *Cosmopolitan. Life* chose Nancy Love as one of the six American women in public life with the most beautiful legs. The War Department Bureau of Public Relations approved of the publicity, but the staff at the Ferrying Division was uncomfortable. It seemed to trivialize the whole ferrying operation. Male ferrying pilots were having trouble explaining why they were moving planes around this country when men were fighting and dying in the skies over Europe or over the Pacific. And this "glamour girls" business—the *Look* article and the *Cosmopolitan* article and Nancy Love's legs—it wasn't helping matters at all.

Colonel William H. Tunner, head of domestic ferrying, saw another problem: The publicity put a lot of pressure on the women. He knew that if something went wrong—and sooner or later something would—then the press which supported the program would turn. They'd call it too dangerous and demand its termination. He wrote a memo to the War Department Bureau of Public Relations, in which he concluded: "Stories of this type are not considered to be for the best interests of the Ferrying Division and will tend to over-glamorize the members of the WAFS in our opinion." He too had doubts that they would even become twenty-five (there were still only ten), and it was too early to tell whether they'd prove, in his phrase, "of permanent value." But the War Department Bureau of Public Relations decided that the attention was on the whole more good than bad—they would only go so far as to say that the subject of women pilots should be treated with "diplomacy and delicacy."

So by November 1942 the WAFS had come to know publicity as a fact of life. The press seemed to give the most attention to Nancy Love. She was running the show, and with those wide blue eyes and Dorothy Lamour cheekbones she was a natural subject for photographs. And of course, there was that name. Nancy was asked to address civic groups and give radio interviews and pose for newspaper photographs. Despite her natural reserve, she would do these things for the good of the program. She thought it would just wear off. But the weeks passed and still the calls came, and Nancy began to be a little bothered. One night at Kent Manor, Nancy and Betty were sitting in Nancy's room. Nancy, obviously frustrated, said, "I don't work for the press. I work for the Ferrying Division." All Betty could say was, "Don't worry, it'll settle down. The attention is only natural, and since the Army is restricting access to most stories, the press doesn't have much to write about." She smiled. "Besides, we're women pilots. We're, you know, glamorous."

Nancy looked in the mirror. "Betty, can I tell you something?"

"Of course."

Nancy pulled the skin below her right eye, the better to observe its bloodshot condition. "Some days I don't feel very glamorous."

7

SNAFU

By the summer of 1942, it had been over a year since Hitler's London Blitz. During those terrible nights in the fall of 1940, when German bombers had hammered London, thousands had been killed, and hundreds of thousands made homeless. Huge chunks of London were still missing. Jackie Cochran would watch the air raids from her blacked-out hotel suite at the Savoy with a grim fascination. By 1942 the English were mostly living shockingly close to subsistence level. Their spirit, however, was still intact. Sometimes it was hard to see it in the tired faces with their tired eyes. But Jackie could still hear it in their voices, and in their language. For the pilots of the Royal Air Force, one measure of altitude was "angels"—one angel being exactly one thousand feet. It was as if the distance from earth to heaven were measured by great invisible seraphim, one standing on the shoulders of the other.

Britain's Royal Air Force (RAF) was so desperate for manpower that it had put aside its qualms about resorting to womanpower almost from the start. Jackie Cochran came to the RAF bearing a gift of twenty-four ferry pilots, and Britain showed its gratitude by treating her like a celebrity. Jackie's ATA recruits were being integrated into the first phase of training at a base outside London, while Jackie herself was being feted by earls and generals. But Floyd was thousands of miles away and she missed him. Besides, things were really happening now in the War Department back in Washington, and it made her more than a little antsy to be so far removed from that kind of action. So Jackie was gratified when, in the summer of 1942, Hap Arnold told her to start thinking about heading stateside. Six months after America's entry into the war, he was beginning to feel that her idea of women in the Air Force was no longer premature.

When Jackie arrived in New York on September 12, *The New York Times* was running a story on the WAFS. This was the first Jackie had heard of Nancy Love's program, and she was furious. The next morning, she was at Arnold's office in Washington, and she had a few matters to discuss with the Commanding General of the Army Air Forces. For two long years she'd been promised the direction of a women pilots program, and now one was approved behind her back and she had to learn about it secondhand. He had told her to see him when she got back from England, and dammit, here she was. And Nancy Love? Brother, you gotta be kidding.

As far as Jackie Cochran was concerned, the Army Air Force had some explaining to do. And, in fact, it did. But the story of the emergence of a women pilots program behind her back was far more complicated than she could have supposed. Military bureaucracy, especially during wartime, is such a dense and intricate organism that even a commanding general like Hap Arnold could occasionally be surprised by it. Which, in fact, was more or less what happened.

When America declared war on Japan in December, it became obvious that airpower would be crucial. Estimates of pilots and aircraft necessary to wage war were multiplied tenfold. In January 1942, General Olds had notified Cochran that he planned to recruit women pilots after all. At this point, Jackie was recruiting American women pilots for the ATA, and she was having difficulty finding women with necessary flying experience and flexibility to commit to an eighteen-month contract overseas. For Jackie, Olds's decision could not have come at a worse time; had he reached it earlier, she could have assumed command of the program, as was her due according to her arrangement with Arnold. But now she was committed to going to England. Worse, if Olds began to recruit from the very small pool of American women pilots with considerable experience, she might be unable to fulfill her promise to the ATA of twenty-five women with three hundred hours each. She wrote to Arnold, observing, "It is terribly confusing to have you say that women won't be needed for many months to come and to have General Olds tell me the next day that there is a shortage of pilots and he is going to use women as a consequence."

Arnold agreed that it was damn confusing, especially because he didn't know the first thing about this plan. Nor did he approve of it; unlike Olds, he did not feel that the Air Force was so short of manpower that it needed women. So he composed a letter to Olds in which he laid down the law pretty plainly: "As per instructions you received from me on this date, you will make no plans or open negotiations for hiring women pilots until Miss Jacqueline Cochran has completed her present agreement with the British authorities and has returned to the U.S." Arnold had made Jackie a promise, and he intended to keep it. That was that. By April, Cochran had found her twenty-five pilots, and was in England overseeing their integration into the ATA.

Meanwhile, in June, Colonel William H. Tunner, head of domestic ferrying, had become desperate for pilots. One day Bob Love, then deputy chief of staff for the Ferrying Division, mentioned to Tunner that his wife was a licensed pilot with more than 1,000 hours. A few days later Colonel Tunner met with Nancy herself, and she assured him that there were dozens more like her. Tunner knew nothing of Cochran's idea, but he was enthusiastic about Nancy's. On June 11 he wrote to his commanding officer, who transferred Love to Washington in order to help Tunner draft a formal proposal.

A few weeks earlier, Tunner's commanding officer would have been Robert Olds. But with the onset of war, the Army Air Forces reorganized itself from top to bottom, a reorganization formalized in early March. By June, Olds had accepted a transfer to command the Second Air Force, and General Harold George occupied the position Olds had vacated as chief of the Air Transport Command. George was unaware of Arnold's letter to Olds insisting that no plans be made for hiring women pilots without his approval and Cochran's participation. Under different circumstances, General George might have waited for approval from Arnold; but in May

1942 Hap Arnold was recovering from a heart attack. So George proceeded on Nancy Love's proposal on his own authority.

Clearly, there had been a snafu; less clear to Jackie—or to anyone else at the time—was its origin. Because neither Jackie nor Arnold was aware of the circumstances under which Nancy Love's plan had been adopted, it looked to Jackie as if Nancy had been sneaking around behind her back, taking advantage of her absence to usurp the position and the program which Arnold had promised her. The consequence was a certain amount of enmity between Cochran and Love. Word of Jackie's anger quickly reached Nancy Love, causing her to fear that Jackie would sabotage her program. For the duration of the war, relations between Jackie and Nancy would be tense. And in the months to come, the press would make much of the rivalry between the two dynamic, ambitious women. Although neither Cochran nor Love would realize it until after the fact, the truth was that the situation resulted from a rather routine breakdown in military communications.

At any rate, by September 1942 Nancy Love's WAFS were already established at New Castle. But Arnold had made a promise to Cochran, and he knew there couldn't be two women pilot organizations. After four days his assistants found a thread by which they could weave Cochran's proposal into Love's program. Cochran had included plans for training. Few women pilots had enough flying hours to qualify as ferry pilots, but hundreds had gone through the Civilian Pilot Training Program, or started flying as a hobby. All could be taught to fly the Army way. So, on September 15, five days after Nancy Love and her WAFS were officially introduced, the Army Air Force announced the formation of a training program to prepare women pilots to serve with Nancy Love's WAFS. It would be called the Women's Flying Training Detachment, and Miss Jacqueline Cochran would be its director. Plans were drawn up for the training of some five hundred women pilots.

In the fall of 1942 the announcement attracted little attention, even with in the AAF. If, as many said, the Air Force was the stepchild of the Army, then the women pilot programs had become the stepchild of the stepchild.

8

HANGAR TALK

On September 21, the WAFS gathered formally in Nancy's office, as a squadron. Nancy was standing behind her desk, and in some indefinable way she seemed more official. She said that while they might have heard rumors that eventually they, like their male counterparts, would ferry pursuits and bombers, perhaps even overseas, these rumors were exactly that—rumors. In fact, they would ferry the smallest trainers, and those only within the United States. She reminded them that some of the men on the base resented their presence. She also told them that they were under a three month probation, at the end of which they would probably be commissioned in the Army Air Force as lieutenants. She had a bit of particularly good news, too. They would be paid almost at once, and their quarters on base were finally ready to accommodate them.

Nancy Love could shift subjects so gracefully, so effortlessly, that a listener might take a minute or two just to notice. And it was a moment before the women realized that she had become gravely serious. She was speaking in even tones, as though practiced. "If the WAFS are to succeed, our personal conduct must be above reproach. There cannot be the faintest breath of scandal.

Among other things, this means you may not accept rides with male pilots." As if to answer the unvoiced question, she continued, "If a male pilot and a WAF were seen leaving a plane together, there would be suspicions that they were playing house in government property." There were sidelong glances about this—not so much concerning the subject as Nancy's way of putting it—"playing house." But Nancy was wearing a thin smile that said she both appreciated the humor of the situation, and knew that a misstep in this area could end the WAFS.

Which was true. The WAFS were very much in the public eye, and Nancy knew that America had yet to make up its mind about women serving as military pilots. The implicit question behind all the newsreels and articles was: What kind of women abandoned their traditional roles to fly airplanes for the Army? Were they courageous and patriotic, or selfish seekers of adventure? Women were already serving by the thousands in all branches of the military, but the majority had been assigned traditional female duties, as secretaries, file clerks, switchboard operators and the like. Only the WAFS had encroached upon what was commonly understood to be an exclusively male domain. If even one opened herself to allegations of scandal, the tide of public sentiment would almost certainly turn against the whole squadron.

Nancy shifted to more immediate matters. Although they were not lieutenants yet, they would have both the privileges and the responsibilities of officers. They would be expected to stand in line for roll call each morning at 8:00 A.M. Colonel Baker, the base commanding officer, had ordered them to participate in reviews. And for that they would need uniforms, which they would purchase at their own expense from a tailor in Wilmington.

Over the next few days the WAFS got their first taste of life in the Army Air Force. After roll call there were classes. They reviewed navigation, and learned ferry routes, ATC procedures, and AAF terminology. But most of all they learned about forms. All AAF planes had a pocket on the right side of the cockpit for Form 1. These were used to list takeoff point, landing point and duration of flight. The other side of Form 1 was Form 1-A. A red diagonal line anywhere on the page meant the plane had a defect. If there was a red *X*, then the plane was unflyable. They would also have to make the final ground check themselves. There was a flight clearance form—a new one would have to be filed at the beginning of each leg—with blanks for name, home base and serial number of the airplane, takeoff point, expected landing point, length of flight, ETD and ETA. The last pieces of information were estimates, requiring that the pilot make a rough calculation of altitude of flight, wind direction and speed. There were forms for weather delay, and there were forms called RONs for "remain overnight," to be filed whenever a mission would require a pilot to be away from New Castle for longer than a day.

The WAFS barracks on base was called Bachelor Officer's Quarters 14. Like most quarters at New Castle, BOQ 14 was a wood frame building. There were two floors, each of which had a long hall running its length with eleven rooms on either side. There were latrines and showers, and one particular feature not included in the standard blueprint for an Army barracks: a ditch, a long trough in the earth out in front. It was quite a formidable ditch, steep-sided and too wide to simply leap across. The women had to wade into the ditch and then scramble up the other side, taking it in turn to give each other a hand. Sometimes the ditch seemed like a reminder that their place in BOQ 14, like their place in the Ferrying Division and the Army Air Force, would have to be earned over and over.

When the women moved into BOQ 14, Nancy and Betty chose rooms on one end of the second floor. They would meet for a drink before dinner. But after hours, when Nancy and Betty would retire, some of the others would gather in Teresa's room, straight across from the bathroom. They knew Teresa was comfortable with them when she told them to call her "Jamesy," and she started calling them "sister" or "old bag." Sometimes she would whisper to Del—almost conspiratorially—"Did you ever think we'd be with such classy dames?" Teresa didn't seem to care for propriety. One night she piped up with, "You old bags know how to jitterbug?" A minute later they were in the hall of BOQ 14 taking dance lessons. Afterward, they gathered back in Teresa's room and collapsed into breathless heaps on the floor, and they talked.

Even when she was sitting on the floor, Del Scharr was a bit self-conscious about her height. When Del was a girl there was a boy down the street who would call her "second-story" and "step ladder." But her mother had told her to be true to herself, and Del

Scharr did not slouch. In truth, however, Del was a bit intimidated by some of the WAFS. They smoked cigarettes and drank liquor, and one of them was actually divorced. Sometimes they did things Del found hard to understand. On the morning the WAFS were issued goatskin flight jackets, Betty Gillies and Helen Mary Clark had rubbed them in the dirt—new goatskin jackets!—to make them look used. What made it most difficult for Del was that they had eastern educations, and sometimes they used words she didn't know. But Del Scharr was nothing if not confident. And as Teresa might say, heck, they're just a bunch of gals like you and me, and I bet they put on their flightsuits one leg at a time just like us.

Still, Del thought, *I've got some catching up to do.*

On those nights in Teresa's room, most of the talk was about flying, and that was a subject Del knew almost as well as anyone there. Crowded together on Teresa's narrow bed and sprawled on the floor, the WAFS swapped stories about their days as barnstormers and flight instructors, recounting first solos, mechanical failures, cross-country trips. Pilots called it hangar talk.

The kind of hangar talk Del was used to took place in airports and usually centered on flying itself. But in BOQ 14, the women often found themselves talking less about flying and more about military politics. The WAFS program was still so new and experimental that many matters of policy had yet to be decided. Regarding the women in its ranks, the Army Air Force seemed to be making up the rules as it went along. For Del, Teresa, and all the others who had left families and careers to join the WAFS, the precariousness of their situation was unsettling. It seemed important to talk over all the possibilities, to gauge their collective feelings about what might await them.

The most obvious topic of discussion in this respect was militarization—their probable commissions as second lieutenants. It would mean greater job security and better benefits—like a clothing allowance for uniforms—and it would represent a level of acceptance. But it would also mean a loss of freedom. As things stood now, the women could leave the program at will. Militarization would commit them for the duration. Some of the women, especially those with small children, had reservations.

The WAFS had other concerns, too. Recently they had heard rumors that Jackie Cochran was starting a training program for women pilots. They knew that upon her return from England, Jackie had been displeased to discover that the Army Air Force had started a women pilots program in her absence, and especially that it had been placed under the direction of Nancy Love. It seemed to follow that as head of a women pilots training program, Cochran would work to undermine the WAFS, or at least make life unpleasant for them.

The WAFS who didn't know Jackie personally knew her by reputation. Del, though, hadn't heard much about Jackie, and some of what she was hearing made her feel that perhaps the others were judging too harshly. To Del, there was something admirable in a woman who started life as an impoverished orphan and transformed herself into one of the hottest pilots in the world. Still, some of the stories about Jackie were so fantastic, so removed from her own experience, that they hardly seemed real. Pat Rhonnie said, "Just guess what Floyd gave the first Mrs. Odlum in the divorce settlement?" Everyone leaned forward a little and Pat paused dramatically and announced, "Bonwit Teller!" Del smiled at the thought of finding oneself the new owner of a New York department store. She said, "What a lovely going-away present!" And everyone laughed.

Del was never comfortable around people with cameras. One day a photographer from *Cosmopolitan* had the WAFS line up, wearing flightsuits and cloth flying helmets. In her flightsuit, Del looked as though she had borrowed a pair of coveralls from an older brother who wore them only on Saturdays to change the oil in the car, and otherwise kept them rolled up in a ball under the sink. Cornelia was standing near Del, and the contrast could hardly have been greater. Cornelia was wearing exactly the same kind of flightsuit, but with her shoulders back, the belt tied casually, and the merest suggestion of curves, the same general-issue clothing looked as though it had come from a collection of the latest fall fashions.

One Sunday afternoon Del was ironing her uniform, and Cornelia stopped just outside the door. To Del's surprise, Cornelia said, "Would you like to go for a walk? It might be nice to catch the leaves before they've all turned." Del agreed. It would be nice to see the leaves. It had been a while since she'd noticed such things.

It was a crisp September afternoon. They walked on the side of a road that went through light woods. Every now and then a car slowed, and an enlisted man asked if they needed a ride. Del and Cornelia smiled and shook their heads and the man smiled back and gave a little salute and the car sped away, swirling a few brown leaves into the air. The two women talked of the people they missed, and where they had been, and what they might do when the war was over. Cornelia told Del a story about her mother—a religious woman who didn't really know much about her daughter's passion for airplanes, except that she didn't like it. "I came home one day, and my mother was in the garden, and I said, 'Mother, I've just soloed,' and, without even looking up from her roses, she said, 'That's wonderful, dear. Now you won't have to do that again.' " They both had a little laugh over that.

It was still early autumn. They walked in silence for a while. The sun was lower now. They were in shadows, and Del's feet were wet. Partly to get her mind off it she asked Cornelia, "Is it true what they're saying? You were at Pearl Harbor?"

Cornelia smiled a little. "Well, I was *near* Pearl Harbor." She paused, and her eyes seemed to focus on a place in the road a little bit ahead of them. "I was an instructor. I was in a two-seater, with a student. And a plane was headed right at us. I thought it was from the air base, so I took the stick from the student, and just dove. And when I looked up through the canopy I saw the rising sun insignia on its wings. I managed to land the plane. I cut the engine and we just jumped out and ran into the nearest hangar. But it was close. A few seconds later the trainer was strafed, just sitting there on the runway." Del couldn't think of anything to say to this. Cornelia looked at her and smiled and said, "If my mother knew I'd been flying on the Sabbath. . . ."

Meanwhile, their training continued. Lieutenant Jordan drilled them every morning in preparation for an upcoming parade review. In a week he determined that they had learned at least the fundamentals, and Nancy herself took his place. She had doubts about her abilities in this area, but her first morning on the field she gave it her best. Walking alongside the ten women, she would pick what seemed like the right moment and try to shout "about face" and "forward march." But it came out hoarse, so she'd swallow and try again. Her soft, husky voice, so effective at close quarters, was ill suited to shouting across a drill field. She began to feel a little self conscious. And self-consciousness made her uncertain about saying anything, and that made her confuse a command, and her confusion in turn made her even more uncertain, and then she hesitated about exactly what to shout next. And the ten marching WAFS, waiting for the halt or the about face that they hoped would come with the next step and then the next, just marched right off the end of the field.

They laughed about it afterward. But it was clear that they'd need to undertake some extraordinary measures if they were to perform well on parade. Betty Gillies, who had learned parade drill at a particularly rigorous finishing school, offered to help. That night they practiced in the hallway of BOQ 14, Betty shouting orders.

Gradually, they improved. Even Nancy grew a little more comfortable, or maybe the women learned to hear her voice. On October 3, the WAFS participated in parade and review for the New Castle commanding officers. They performed commendably. When the drill was complete they stood at attention. For the first time they were wearing their uniforms in an official capacity. They were gray serge jackets with padded shoulders and patch pockets, and matching slacks. On the left sleeve of the jacket was the arm patch of the Air Transport Command, and over their left pockets were the wings of ATC Service Pilots. Helen Mary and Del Scharr and Teresa James and the rest all looked resplendent. But Cornelia seemed to be holding her head a little straighter, and her gaze seemed a little clearer. On that October afternoon, as the WAFS were standing at review, it happened that a B-17 and four pursuits took off from the base and flew east—headed out across the Atlantic. As Cornelia stood at attention, tears came to her eyes.

Later, she would write about this moment. Cornelia had set out to become the unofficial WAF historian and poet. It had occurred to her that in this place and time, two dramas were being played out simultaneously. The first was the ferrying itself—the marshaling of forces against the clearest and most horrific incarnation of evil in the twentieth century. The second was the beginning of American women acting as military pilots, the participation of half the human species in the new Air Age. If Cornelia did not know which drama would be remembered in a thousand years, she did know that there could have been no *Iliad* without a Homer. And she did know that she had been provided an education, the skills of a wordsmith and, most important, the good fortune to be present. The WAFS, Cornelia was sure, were about to perform something astonishing. And she was determined to become their memory.

9

THE WAFS

"Ladies, if I could have your undivided attention for a moment." Nancy was sitting at the head of the table in the BOQ 14 meeting room. The ten women were crowded around her, some sitting, some leaning against the back wall. Nancy straightened the pad in front of her, then glanced quickly around, and began. "Formation flying involves set practices, and we must understand them. The flight leader decides who will navigate each leg. She judges the navigating. The navigator takes off first." She put a large *X* in the center of the paper. "Then the pilot assigned the number-two position takes off, and moves to the navigator's right." She put a second *X* to the right and behind the first. "The third is behind the others and to the left." She made a third *X*. "The flight leader flies in the rear, as she has responsibility for all other pilots. When the navigator is ten miles from the destination airport, she'll throttle back a little so the flight leader can come forward to lead the group in. At that point, you'll reposition yourselves into echelon formation."

"Echelon formation?" Del was speaking for several of them.

Nancy turned the paper over. She made a series of *X*'s in a single diagonal, each to the right and behind the one before it.

Betty said, "Don't you think we should have some signals? If a girl has engine trouble she could dip her nose a few times before she goes down."

Nancy nodded. "Good. The flight leader watches all of you. You watch each other, and you watch the flight leader, too. If the navigator is off-course the flight leader should come forward and take over."

Del said, "Maybe we could dip our right wing, if, for instance, the navigator is headed too far left."

"A sensible solution. Let's do it."

Teresa said, "And lifting the nose might mean we should climb to a higher altitude? We might find it easier to cruise. . . ."

Nancy was resisting a smile. Already they were anticipating problems, and solving them. And already, they were watching out for each other.

She reviewed preflight procedures, some regulations about RONs, and gave some last minute advice. "Stay at least five hundred feet from each other. Stay beneath and to the side of clouds. And stay high enough over towns that you can deadstick outside them in an emergency." No one in the room needed this advice; it was like telling them to wear a sweater because it might get cool. When Nancy stopped there was an awkward pause. Anyone could see that she was like the mother at the front door sending her child off to the first day of school, suddenly feeling a bit irrelevant. Betty grinned and decided to put an end to it. "Anything else, Nance? Like 'Make me proud?' " Nancy replied quietly, "You already have." And then she grinned a little, too. "Good luck, ladies."

The next morning on the bulletin board at ops were orders for the delivery of six L4-Bs from the Piper Cub factory in Lock Haven, Pennsylvania, to Mitchell Field on Long Island. L4-B was the AAF designation for Piper Cubs—the *L* stood for "liaison." They were single engine, sixty-five horsepower planes. The whole flight would take only about eighty minutes, but it was the first mission for the Women's Auxiliary Ferrying Squadron. And it was the first mission for assigned ferrying pilots Rhonie, Fort, Clark, Scharr, James and Gillies.

On the morning of October 22, the six women in flight coveralls carried their parachutes out of a Boeing twin-engine, which had just arrived at the airfield of the Piper Cub factory in Lock Haven. There were stares from ground crews and other pilots but Betty and Del and Helen Mary and the others just affected an air of normalcy. They found the planes whose tail numbers were in their orders, inspected them, climbed in, waited for the line crews to crank the props, and received clearance from the tower. One by one, the six little trainers became airborne.

As designated navigator, Pat had a series of maps called sectionals, pleated like an accordion and tied to her leg with a string so they wouldn't get blown out of the cockpit. On the sectionals she had drawn a line from origin to destination. The line passed between checkpoints indicated on the map—transmission lines, a watertower, the bend in a river, a rail line, a mountain ridge. Sitting in the open cockpit, with the map tied to her leg, she watched as the low wooded hills of the eastern Alleghenies slid by beneath her. She would expect to see a watertower on the left, ten miles later a ridge, and then know that she must be approaching a town like Berwick or Hazelton.

The little trainers were light and underpowered. When they were bucking a headwind the women could look down and see cars and trucks on the highways actually moving faster. And so even those two hundred miles were a long flight. But eighty-three minutes after the six trainers had left Lock Haven, they were on the final leg of the landing pattern at Mitchell Field. They were on schedule. It was a perfect delivery.

Meanwhile, in New Castle, rooms at BOQ 14 were filling quickly after all. A flight instructor named Nancy Batson had not received the telegram. Instead, she'd read about the WAFS in a Birmingham, Alabama, paper. She simply bought a train ticket to Wilmington, then took a cab to the base. A woman named Gertrude Meserve had arrived from Boston. Barbara Erickson had taken the train from Walla Walla, Washington, where she had quit her job as a flight instructor. There was a swimmer named Katherine Rawls Thompson who had won two gold medals in the 1936 Olympics. A flight instructor named Evelyn Sharp had almost 3,000 hours of flying time. And there was a small, shy woman named Bernice Batten who had been unable to afford transportation to New Castle, and had hitchhiked most of the way from west

Texas. By November 1942 the WAFS numbered twenty-five. As far as Betty and Cornelia and Del and the rest were concerned, Nancy Love had performed a small miracle.

Soon all of the women were ferrying regularly. Each had been issued a large canvas sack that was cross-stitched for strength. It was called a B-4 bag, and it carried winter flying suits and a navigation case. It was not unusual for a fully packed bag to weigh ninety pounds. They would carry and drag and push the bags along with their packed parachutes, getting a ride to the Piper Cub factory at Lock Haven via train or plane. Then they would find the planes whose tail numbers matched those on their orders and begin a preflight check.

On December 11, Del Scharr, Helen Mary Clarke, Esther Nelson, and Gertrude Meserve were to fly five L4-Bs from Lock Haven to Abilene, Texas. Del was named flight leader. Each leg of the trip would be sixty to ninety miles long, and they'd make about five legs a day. Because Ferrying Division regulations prohibited them from flying after sunset, each evening they landed, tied down the planes, and used the seat belt to tie the stick so that the wing surfaces wouldn't move. Then they closed the flight plans, watched the line boys fill the tanks, checked parachutes in lockers at operations, and found a Western Union office to send RONs back to New Castle. On all overnights, there was the question of a place to sleep. If they were lucky there were nurses' quarters or maybe WAACs barracks. Sometimes they had to find a hotel in town.

One morning when Del called the local base she learned that the whole of the Ohio Valley was restricted to instrument flight. The L4-Bs had no navigation instruments. It was two days before the weather cleared. When it did they borrowed a broom from some farmers, and swept the snow off the wings. The five Cubs climbed into the cold air and headed west again, across the rolling hills of the Ohio Valley to Columbus, then over the Indiana border to Richmond, on to Indianapolis, Terre Haute, St. Elmo, Illinois, Scott Field, southwest over the Ozarks, across Missouri and the Oklahoma plains. Then seventy-six miles to Tulsa. Near Tulsa a low fog caused them to lose a day. Then it was on to Oklahoma City and across the Red River Valley. Finally, eleven days after they'd left New Castle, they reached Abilene.

The return trips could be almost as exhausting as the missions.

Sometimes they flew back to New Castle on a commercial airliner; sometimes they went by bus. Most often they took the train. In 1942 trains were on wartime schedules, and so they were crowded with soldiers on passes and civilians visiting servicemen. Although the larger train stations employed baggage carriers, there was little incentive for them to carry heavier baggage, and when WAFS appeared with ninety pounds of it, most redcaps quickly looked in another direction. So when WAFS returned together, they worked out a system. One would get in line at the ticket office, while the others would carry, drag and push the bags onto the platform. When they finally got on the train they would have to put their B-4 bags and parachutes in the aisle. Sometimes, the trains were so crowded that no seats were available. But the women were so bone-weary that there in the aisle of a crowded railcar, with other passengers pushing past them, they would sit on their B-4 bags, lean against an armrest, and simply fall asleep.

10

THE GUINEA PIGS

The telegrams went out on November 9, 1942. Thirty prospective trainees were invited by the Army Air Force to report to Houston, at their own expense. If they wished to pursue EMPLOYMENT ON CIVIL SERVICE STATUS AT THE RATE OF ONE HUNDRED AND FIFTY DOLLARS PER MONTH DURING . . . SATISFACTORY PURSUANCE OF FLYING INSTRUCTIONS UNDER ARMY CONTROL, they were to present themselves at Houston's Rice Hotel by 10:00 A.M. on November 15.

Since her September meeting with Hap Arnold, Jackie Cochran had been busy scouting out locations, hiring staff, and reviewing the records of hundreds of women pilots. All of this had been more difficult than she had anticipated, but things were finally in place. The telegrams went out, and the Army Air Force training program for women pilots—the program Jackie had worked to begin for almost two years—was finally under way. Jackie was director of the 319th Army Air Forces Flying Training Detachment, also called the "Women's Flying Training Detachment" (WFTD). Each class would have a sequential designation. The women who responded to the telegrams would be "43-W-1"—the first class of women to graduate in 1943.

It was a small beginning—thirty trainees, makeshift facilities, and, if truth be told, a somewhat haphazard plan for the training program itself. Jackie had wanted better for her girls. She had wanted better planes, better facilities, better weather. Still, she was confident that this was the start of a major contribution to the war effort. If things went well, this small group of trainees would pave the way for hundreds, maybe even thousands of women to serve as ferry pilots for the Army Air Force.

Before the war, the Rice Hotel had been Houston's emblem of gracious southern hospitality and quiet refinement, its lobby populated chiefly by decorous society matrons and liquid-lunching businessmen with slow southern drawls. But by 1942 the same lobby was overrun with swarms of bustling men in uniforms. The sleepy southern city had been jolted almost overnight into the frenetic pace of wartime activity. The people of Houston did their best to get used to the Army's invasion of their Rice Hotel, and in time they nearly succeeded.

All the same, the lobby of the Rice Hotel was a strange sight on the morning of November 15. That morning, twenty-nine women stood together in the lobby and they were quite an assortment. One had marched in trailing two Afghan hounds and ten trunks of clothes, and she was dressed about the way a woman with ten trunks of clothes might be expected to dress. Three or four of the others gave off an unmistakable aura of East Coast old money. Another was wearing trousers and cowboy boots. Beside her stood one who, in her neat, plain skirt and well-pressed blouse, might have been a University of Nebraska coed. Beside the coed was one who seemed to be a skilled practitioner of what Houston's elderly society matrons would call "face painting." There was a reason these women looked like such an odd assortment. They *were* an odd assortment. Their occupations ranged from actress to nurse to dancer to stuntwoman to homemaker to blackjack dealer to air traffic controller.

Then things got even stranger, because as the women stood patiently waiting, a phalanx of military men surged through the front door and began to march across the lobby. Phalanxes of military men in the lobby of the Rice Hotel were nothing new in the fall of 1942, but this phalanx was headed up by a woman. On the lapel of her impeccably tailored suit was a large pin in the shape of a propeller. It had a diamond in the center.

Jackie Cochran welcomed WFTD Class 43-W-1 to Houston, and said, "You have the honor and distinction of being the first women to be trained by the Army Air Force. You are very badly needed." Jackie told the trainees that they would report to the airfield for their first classes the next morning. In the meantime, there was one thing they had to get straight. At least for now, the Women's Flying Training Detachment was secret. The Army Air Force regarded them as an experiment, and thought it best to avoid all publicity unless and until they succeeded. The women of the WFTD were required to have only 75 flying hours, and there was no telling whether hundreds of them would be able to fly the Army way. Jackie told them that if anybody asked them what they were doing in Houston, they could say anything they liked, as long as it wasn't training to fly airplanes for the Army Air Force.

She then introduced them to Leni Leoti "Dedie" Deaton. Mrs. Deaton, she informed them, would be their Establishment Officer. Thirty-nine years old and a former national field volunteer for the Red Cross, Mrs. Deaton wasn't quite sure what an Establishment Officer was. All Jackie had told her was that she was to "look after the girls." Because there was no housing available on the airfield, her first task was to find them lodgings. She had arrived at the Rice Hotel with a list of local rooming houses, only to find that many of the trainees had already arranged for accommodations on their own, which discovery only served to reinforce the feeling she was already getting, just looking at them. These weren't "girls," they were women. And sophisticated women at that. Mrs. Deaton, who had never left Texas in her life, was suddenly beginning to feel overwhelmed by her new job.

One of the officers who accompanied Jackie to the Rice Hotel was a young lieutenant named Alfred Fleishman. Many AAF officers were openly hostile to admitting women to their ranks, but Fleishman was more open-minded. After the meeting in the hotel lobby, the lieutenant approached Mrs. Deaton and offered her his copy of *The Officers' Guide and Army Regulations*. If the women trainees were to be successfully integrated into the Army Air Force, he suggested, it might be a good idea to start acclimating them to the Army way of life. Mrs. Deaton accepted the manual gratefully. That night she read it cover to cover, and decided that this was how things would have to be run. There would be rules and there would be order.

There remained problems of logistics. The WFTD would train on space belonging to Aviation Enterprises at Houston Municipal Airport's Howard Hughes Field. Because the women were scattered at rooming houses and hotels all over town, Mrs. Deaton needed to find reliable transportation to and from the airfield. Aviation Enterprises offered to lease her a bus. It had been used for a few years, but it was, they assured her, in good working order. The same day, she found a local jack-of-all-trades with the improbable name Tex Thrash, who agreed to drive the bus to and from the airport.

At 6:00 A.M. on the morning of November 16, a group of trainees waited outside their hotel for their transportation to the first day of classes. When it arrived, they were astonished. It was a large school bus, painted white, with little striped awnings over the windows and red edelweiss painted along the sides. These adornments were explained by the flowing script that ran across the back: "Tyrolean Orchestra." Aviation Enterprises had indeed acquired the bus secondhand, and had not bothered to modify its decor. The bemused trainees climbed on board. And Tex Thrash ground his gears and rumbled out onto the highway.

Half an hour later, the "Tyrolean Orchestra" rolled through the front gate at Houston Municipal airport, then made its way down a long dirt road. Soon the women got their first look at the place where they'd learn to fly the Army way. Howard Hughes Field boasted a hangar, an equipment shed, a runway, and not much else. It had no housing facilities, no dining facilities, no classrooms, no pilot's ready-rooms, and exactly one rest room. The nearest food was in the cafeteria in the airport's main terminal, a half mile away. In November, the height of Houston's rainy season, this half mile was a half mile of mud.

The airplanes themselves were twenty-two different kinds of small trainers. All had seen better days. And then there were the *other* airplanes. Howard Hughes Field lay adjacent to Ellington Field, where the Army Air Force was putting thousands of combat pilots through their paces in everything from basic trainers to

B-17s. The airspace above Houston Municipal was dangerously crowded in the fall of 1942.

At 7:45 the thirty women took their seats in the makeshift classroom where, in the weeks ahead, they would log some 200 hours of ground school. Mrs. Deaton introduced the chief instructor, Bert Black, and he introduced 43-W-1 to the world of Army aviation. "The first thing you'll have to do," he said, "is forget everything you thought you knew about how to fly. Forget it now. Don't expect your instructor to be polite and gentle. He won't be. And if he cusses at you, you probably deserve it, so don't go running to Mrs. Deaton to complain."

The women, acknowledging their experimental status, had begun to call themselves the Guinea Pigs. The Guinea Pigs learned that they would be putting in twelve-hour days at Howard Hughes Field, and that they would be taking homework with them when they left. They were warned that their schedules might not permit time for the midday trek to the main terminal and back, and that this meant they might sometimes have nothing to eat between dawn and dusk. As their training progressed, they found that their instructors were indeed quick to criticize and slow to praise. They also discovered that some of these instructors had less experience than they did. They were informed that the weather they were experiencing now—fog, rain, sleet and mud—would last through December, and January, too. They learned that their short breaks would often be spent waiting in line to use the facility's one toilet. They learned to march. They learned to be tired, hungry, muddy and cold. And of course, they learned to fly the Army way.

WFTD flight training was a grueling course, designed to turn fledgling aviators into military transport pilots in a matter of months. Primary training required twenty hours in the cockpits of the PT-17 and PT-19, and involved elementary maneuvers like loops and rolls. The next stage was basic training, which taught the military applications of techniques learned in primary. Trainees spent seventy hours in 450–horsepower BT-13s, planes slightly larger than the primary trainers, with enclosed cockpits, and radios. They practiced aerobatics, formation flying, and daylight navigation. Finally, in advanced training, they flew AT-6s, and took up instrument flying and nighttime navigation. They made cross-country trips, sometimes as far as California, a challenging route over long stretches of empty desert and high mountain passes.

Each stage involved several hundred hours of ground school classes in aerodynamics, instruments, engine maintenance, meteorology, and navigation. All classes had tests, both written and practical. If a student failed to master any part of the process, she was washed out.

The East Coast heiresses tried not to think what their families would say if they could see their tangled hair and mud-spattered faces. They looked in the mirror at the end of the day and all they could do was laugh. A few women washed out and a few were discharged for medical reasons, but none of them quit. The Guinea Pigs had been told that the future of women in the Army Air Force rested on their shoulders. And Lieutenant Fleishman taught them a simple mantra: "If the Army can dish it out, I can take it." The Army dished it out fast and furious at Howard Hughes Field. And the Guinea Pigs took it.

Meanwhile, the second class was scheduled to arrive in Houston by mid-December, to be followed by another new class every thirty days. Mrs. Deaton was trying desperately to find more lodgings, but few establishments were willing to board young women coming to town to engage in unspecified activities. Finally she found a little row of one-room houses with an office at one end, a type of lodging termed a tourist court. Because tourist courts were inexpensive and because guests did not have to walk through a lobby to reach their rooms, they were known as a favored site for extramarital assignations. So when Mrs. Deaton called Jackie Cochran, who was back east interviewing prospective trainees, and proposed to house some of class 43-W-2 at a place called Oleander Court, Jackie would not hear of it. Mrs. Deaton said there was no choice. On December 8, she received a telegram from Jackie, who had reconsidered. The use of tourist courts was approved. A short-term lease was recommended.

The newer classes pronounced "WFTDs" as "Woofteds." If they had a name, they still had no flight suits and no uniforms, and they arrived at Howard Hughes Field each morning in a motley assortment of outfits. As winter approached and the air in the open cockpits got colder, they were wearing more and more of their wardrobes on their backs. Their clothes were becoming tattered and threadbare, and their loafers and oxfords weren't standing up to the Texas mud. Lieutenant Fleishman managed to commandeer some thick military sweaters, which at least helped take the edge off the incessant wind. Finally, in mid-December, a shipment of actual flight suits arrived. The women gathered around as they were unpacked and distributed, and then they held them up and tried them on and burst out laughing. The flight suits were, of course, designed for men, but this particular batch had been designed for especially large men, and the seats hung down to their knees. Someone suggested nipping the legs in at the bottom. A few of them tried it, and they all started laughing again, because now the flight suits bore a striking resemblance to the baggy pants and generous cuts favored by fashionable gentlemen in Harlem. The women began to call them zoot suits.

Texans dealt with all the mud by wearing cowboy boots, and the Woofteds decided to follow their lead. Before long, Howard Hughes Field had become the site of a virtual revolution in aviation attire. Zoot suits and cowboy boots. It got even stranger when some of the instructors complained that the girls' hair whipped around when they were aloft in those open cockpits, and word came down that they were to wear hairnets.

At night, when the Tyrolean bus had brought them back to Oleander Court and the rooming houses, they would crowd together in one of the rooms and flop down on the lumpy beds and kick off their cowboy boots and just talk. They did not talk about the bad food or the incompetent instructors or the achingly cold weather. They did not talk about their families back home or their husbands and boyfriends or their plans for the future. They talked about flying—about PT-19s and BT-13s and spins and stalls and lazy-eights. The hair nets and the mud and the calisthenics and all the rest of it slid into perspective as nothing more than minor irritations. Because it was worth putting up with almost anything to spend a few hours of each day in the air.

Meanwhile, the people of Houston weren't exactly sure what was going on at Howard Hughes Field. Suddenly it was as though somebody had put a women's college out there. As for the trainees, they knew they were supposed to be a military secret, but they weren't sure exactly how they should answer if they were asked who or what they were, and so they said a lot of things. And one day for no particular reason one of them said, "We're a women's basketball team." The other girls thought that was a real hoot, especially because a lot of them were under five feet two inches. But that became the standard answer. Whenever she was asked what brought her to Houston, a Woofted would smile and say, "Well, I'm a member of a women's basketball team."

And the person who asked her the question would have no idea that she'd spent the better part of the day in an AT-6 Advanced Trainer at 5,000 feet practicing shallow dives and snap rolls.

II

CHRISTMAS, 1942

Back at New Castle, BOQ 14 was mostly empty. Del Scharr and Teresa James would arrive, tired from the train ride, and pull and push their B-4 bags and parachutes over the plank and into the first floor hallway. It was quiet, and that was a little strange. A few weeks ago it had been as noisy as a college dormitory. Now it seemed a little lonely. After she unpacked, Del wrote long letters to her husband. She wrote that since her return, Nancy had given several of the women assignments as squadron leaders at other ferrying bases. She wrote that when Cochran's first class graduated—in early spring if all went well—the WAFS would become their tutors and commanding officers. Del herself had been assigned WAFS squadron leader at the Third Ferrying Group at Romulus Air Base near Detroit, which meant that she would probably have more chances to see him. Still, it was a little sad there in BOQ 14.

Two days before Christmas, Del was writing letters again when Cornelia appeared in her door. Her face was white, and she stuttered her words. She had just gotten off the phone. Fortland, her family home, had burned to the ground.

There was silence, and Del waited.

"It's my journals. When we learned of the transfers I sent them all home; they were in my bedroom."

Del said, "But you can write them again. What's important is that no one was hurt. Your mother?"

Cornelia smiled a little. "My mother's fine," she said.

Then Cornelia seemed to remember her manners, and realize that all this was an imposition. She changed the subject, and asked what Del had ferried lately, and who was where. Gradually, Cornelia and Del began to talk of other things—the war, friends, family, and Christmases past.

On January 27, the Eighth Air Force struck the U-boat construction yards in Wilhelmshaven at the mouth of the Weser River. Fifty-three bombers hit the installation; three were lost. It was the first American incursion into German territory. By the beginning of 1943, the demands of Operation Torch, the Allied invasion of French Northwest Africa, had left the Eighth Air Force weakened and able to mount no more than a hundred bombers for any one raid on the Continent. In the United States, aircraft manufacturing plants were operating around the clock.

Almost from the beginning of the program Nancy had made repeated requests to Colonel Tunner that the women be allowed to "transition"—that is, train to fly other planes, advanced trainers and even pursuits like the P-51 and P-47. Nancy understood that the Woofted training program included an advanced trainer called the AT-6, and because the WAFS were to become the Woofteds' teachers, it seemed only reasonable that they stay a step ahead. But the head of domestic ferrying denied her requests. Nancy knew that as director of Women Pilots she could transfer herself to any base that had accepted WAFS, including those with a more open-minded environment. And she did. Leaving Betty Gillies as executive officer at New Castle, Nancy relocated to Dallas in early January 1943, where she began training with the Fifth Ferrying Group; in a matter of weeks she had transitioned through advanced trainers. To fly other aircraft she'd have to move again, so in late February she transferred herself to the Sixth Ferrying Group in

Long Beach, California. Fellow WAFS B. J. Erickson, Bernice Batten, and Cornelia Fort were already there nearby. Were some of the nation's largest aircraft manufacturers. North American Aviation was among them. And so it came that Nancy Love transitioned into the plane many were calling the hottest in the world.

In 1940 the Royal Air Force had contracted North American to design a new pursuit. Specifically, it was designed as an escort for the bombing raids of the B-17. It was fast, with a top speed of 430 miles per hour, and amazingly maneuverable. North American had called it an Apache, but the Royal Air Force had a better name—Mustang. In time it would be called the best pursuit in World War II, but the first model gave some trouble. As a class, pursuits had three times the accident rate of heavy bombers, and at least at first, the P-51 had its own problem—a nasty habit of "occasional engine failure" on take off. To make the engine run smoothly, trainees were told to leave the runway at full throttle—but even then it sometimes stalled and had to be restarted in midair. Sometimes the engine actually caught fire. Loss of aircraft in actual combat was called an "operational" loss, and loss in any other capacity—training, ferrying and so on—was a "nonoperational" loss. There was a 1943 report called "RAF Mustang Losses Due to Non-Operational Causes." To read it gave one pause: "Destroyed in Forced Landing at Burrough Grave Field, Bucks . . . Flew into high ground in fog at Kimmeredge, Dorset . . . Crashed after engine failure at Burwell, Cambs . . . Crashed near Poppleton, Yorkshire, after engine failure . . . Dived into ground near Bellingham, Northumberland . . . Flew into high ground in cloud near Hillcott, Wiltshire . . . Crashed after low-altitude engine failure near Ockley, Surrey . . . Flew into high ground in cloud at Hindhead, Surrey . . . Crashed on landing at Odiham after collision with Spitfire . . . Hit by AG376 after landing at Dunsfold . . . Exploded in mid-air near Boyttisham . . . Caught fire in the air and crashed near Ringway . . ." The P-51 flight manual had a long section on handling emergencies.

But the P-51B Mustang was a lean and powerful machine, and one of the most streamlined aircraft ever built. It began with an eleven-foot Hamilton Standard propeller. Behind that was an enormous cowling that held a 1,490-horsepower Packard Merlin engine. The Plexiglas cockpit canopy rose behind, and beneath the cockpit was the airscoop, and then the belly just curved up into the place where the fuselage narrowed until, a few feet later, it flared into the tail. Anybody seeing the P-51 sitting out on an airfield with morning sun glinting on its wings would call it beautiful. A pilot would feel the pulse beat a little faster.

One clear winter day in late February, Nancy walked out on the field at Long Beach Army Air Base—cloth helmet tight over her head, her goggles on her forehead. The WAFS had been flying for months now. The photographers and newsreel people weren't there as much, but there were still reporters from some local paper, and the ever present looks from another bomber crew, or some mechanic. But when she saw what she was about to fly, she forgot all about any audience. Its nose rose above its tail, and it looked as though it were crouching, ready to fairly leap into the air.

She did a walkaround. Then she climbed up on the wing, held the canopy and stepped into the cockpit one leg at a time. Resting a foot lightly on the edge of the seat, then holding the edge of the canopy with both hands, she slid down. With her feet she tested the rudder pedals a little and adjusted them slightly. She raised the seat an inch or two, locked her shoulder harness, reached up behind her head and slid the canopy forward and shut. She started the engine and then spoke loudly into the microphone. "Tower, this is P-51 2106462 on runway three requesting clearance for takeoff." Even with the canopy closed she could barely hear herself over the engine. There was a second of static, and she heard in the earphones: "P-51 2106462, this is Tower. You are cleared for takeoff."

They were right to call it mustang. It felt like a living thing, an animal—all glaring teeth and lean muscle. She had been holding it back, and she was about to let it have its way. She gently pushed the throttle until the pressure gauge read 61 inches and the tachometer 3,000 rpm. Now the engine was a roar and she was moving down the runway and gaining speed—*45*, *60*—and the rear wheel lifted until the tail was level with the nose, and for the first time she could see straight ahead—*75*, *80*—at 100 miles per hour she pulled gently on the stick and held her breath for that moment when the plane actually lifted itself from the runway, and the ground fell away. Through the forward section of the canopy she could see only blue sky. It was a magnificent day for flying.

Nancy had a sort of friendly transition competition with her aviator husband, and that night she wrote him a teasing letter, saying, "Just try and keep up with me . . ." And he'd write back, "I know better than to try . . ."

A few weeks later Nancy was delivering a Mustang. In the first few months of 1943 the B model P-51 was still new, and so when the control tower operator saw one circling his field he just watched it. But the pilot was maintaining radio silence, just circling. In his headset, through the static, the control tower operator heard a woman's voice: "Okay to come in now?" But he couldn't see her—all he could see was the P-51 and it was still flying the standard holding pattern. So he responded to the voice: "Where are you? Can't spot you at all." The P-51 banked and turned toward the tower. All of a sudden it was *below* the controller, and it flew so close it rattled the window frames. The controller heard the woman's voice again—this time it was almost sweet. "*Okay?*" And the next sound was, "Uh, P-51 cleared on runway four . . ."

It was spring 1943. Nancy Love was one of the first of a handful of pilots to test and deliver a P-51B. It would be several months before AAF pilots would use the plane in combat.

In late March Del Scharr delivered an L-5 to Alamo Field. She was in an airline ticket office, booking a return to Romulus when a clerk, seeing her uniform, registered a kind of recognition. "Did you know Cornelia Fort?" She said it cautiously, but almost casually, as one might ask a stranger on the street if he has the time. The woman was holding a folded newspaper toward Del. "She got killed nearby here. It's all in here." Del felt the muscles in her throat tighten. Slowly, she stood and walked to the desk. Then she took the newspaper from the outstretched hand and tried to focus her eyes.

When Nancy Love received the news in Long Beach, she was not surprised that it had happened. It would have happened sooner or later. But when she read the AAF accident description, she was just a little angry at how it had happened.

> It appears that the landing gear leg on F/O Stamme's aircraft No. 42450 broke off the left wing tip of Pilot Fort's aircraft No. 42432, and peeled approximately 6 feet of the leading edge of the wing toward the center of the aircraft . . . F/O Stamme's craft did not go out of control. Pilot Fort's aircraft . . . went into an inverted dive, slowly rotating to the left. There was apparently no attempt to recover or to use the parachute . . . the aircraft apparently hit vertically and did not move from the point of contact.

March 21 was a warm, clear spring day in Texas, and seven BT-13s were enroute from Long Beach to Dallas. During a refueling stop a pilot named Frank Stamme began flirting with Cornelia. When they were airborne again he began to put his AT-6 into dives right past her. It wasn't funny. He was no fighter ace, and he had little or no training in formation flying. But he circled and came by again . . . even closer. And Cornelia did what Cornelia always did. She played it by the book, just maintaining loose formation. She must have hoped he would stop soon, but he did not stop. He just kept diving past her. And on one such dive his wheels were too close. They tore off a wingtip, and Cornelia's plane immediately lost all control. She went into a spin and flew straight into the hard west Texas earth.

When recovery crews reached the wreckage, they found the engine buried two feet in the ground. The switch was off. She had had time to cut power to the engine. It was clear that she knew she was going to crash, and with the power cut she might prevent an explosion and fire. Of Cornelia herself, little could be identified with certainty, but they did find a short string of cultured pearls.

One of the most moving passages in Homer's *Iliad* occurs when the war is over and an old, defeated Priam crosses the smoking battlefield to beg one favor of the victorious Achilles—that the bruised body of his son Hector be returned to him so that he may bury it. The Army Air Force of 1943 understood something of Priam. Male pilots in the Ferrying Command were entitled to a ten thousand dollar insurance policy, and a military funeral with a flag-draped casket. The grieving family was entitled to display the Gold Star. But because the Women's Auxiliary Ferrying Squadron had yet to be made part of the Air Force, Cornelia Fort had a right only to two hundred dollars and a pine box.

The Fort family, however, had tremendous resources, and at the Christ Episcopal Church in Nashville there was a large funeral. Nancy Love and Betty Gillies attended. It seemed everyone knew the Forts. The governor himself sent a bouquet of yellow roses, and the altar was full of flowers. Nancy Love was both sad and angry, but she knew that this was not the time to voice her feelings. Later, she wrote a letter to Mrs. Fort—the religious woman who loved her roses and could never understand her daughter's passion for airplanes.

Much earlier, a few days after Pearl Harbor, Cornelia had begun to consider the killing that she had seen, and thought to comfort her mother in the event of her own death. She wrote her mother the kind of letter soldiers write on the eve of a great battle, a letter in which she tried to comfort those grieving for her.

I was happiest in the sky
at dawn when the quietness of the air was like a caress,
when the noon sun beat down
and at dusk
when the sky was drenched with the fading light.

Think of me there and remember me,
I hope as I shall you,
with love.

What nobody said out loud was that this death from hassling, this death from somebody performing a stunt, was a death no one deserved. And most especially, it was not a death befitting the earnest and literate woman who was Cornelia Fort. Lieutenant Stamme was reprimanded, but he was not court-martialed, simply because ferrying pilots were needed so desperately. There is no record of Stamme's feelings regarding Cornelia Fort's death. Probably he did not know that Cornelia had kept a journal of her days at New Castle and Long Beach, and that she intended to write a history of the WAFS. So although Stamme must have known that by casual disregard for AAF procedures, he had killed one of the WAFS, he could not have known that he had also killed their memory.

12

"KID, YOU NEVER HAD IT SO GOOD"

A few days later Del returned to Romulus. Nancy had appointed Del squadron leader, and Del was proud to have been deemed worthy of the responsibilities that came with the position. But sometimes her responsibilities were a little odd. On March 29 she called her squadron together and made an announcement: "I'm, uh, supposed to find out from you when your next period is coming up and inform the flight surgeon." There was quiet for a moment. Then Barbara Poole muttered, "That's no one's business but my own," and the others looked both baffled and angry. Del sighed because she now had to explain the reason for this directive, and the reason was going to make it even worse. "We've been ordered not to fly during our menstrual cycles," she said, explaining that Ferrying Division administrators had found a passage in the CAA handbook that suggested women often fainted under the duress of menses.*

*Del did not know that the directive was also a response to Cornelia Fort's death, although the official investigation showed her blameless.

Del had never noticed any difference in her stamina or her alertness from one time of the month to another, and she thought this was about the most harebrained regulation the Army had come up with yet. But she had a duty to enforce it, so she took a deep breath and repeated her question.

But it seemed that not a single member of the squadron had any idea when this event might take place. It seemed, in fact, that every WAFS at Romulus had a hopelessly irregular and unpredictable cycle. The flight surgeon couldn't bring himself to question WAFS Scharr about her remarkably ambiguous data, and he couldn't imagine trying to strong-arm these women into providing him with more precise accounts. As a result, no WAFS were grounded at Romulus. The Air Transport Command surgeon noted that any such regulation would be impossible to enforce. The chief surgeon of the Army Air Force wanted to know how they were dealing with the issue down in Houston. And the base commanders at Sweetwater hadn't realized that it needed dealing with at all. After several weeks it was decided that if a pilot chose to ground herself at certain times of the month, she would not be penalized. Otherwise, the AAF did not, after all, need to concern itself with its female pilots' reproductive health.

Soon the women pilots at Romulus decided that "that time of the month" merited its own military designation. In a pun on "Kotex" credited to various feminine wits, they dubbed it "Code X."

Most Army Air Force pilots were male, and the Air Force didn't change its procedures because it now employed a few dozen women. The bureaucratic oversights that resulted made for unusual situations. For Del, it had started back at New Castle. The WAFS were not going to fly combat missions, and so they were unlikely to be exposed to chemical weaponry. But because all military pilots were required to undergo a course in chemical warfare, Del and Betty and the others all learned to put on a gas mask, and to recognize various poisonous gases. For a brief moment each experienced the effects of tear gas on the central nervous system. Once she got over the coughing and the crying, Del had been fairly annoyed. It seemed a ridiculous waste of her time and the Army's money. But Del was new to the ATC back then. By the spring of 1943 she was starting to acclimate. She had sat through classes on how to survive in the Arctic and in the jungle and had attended a film on the effects of social diseases on male genitalia.

If the Army Air Force sometimes ignored the women in its midst, many of its male pilots did not. One night Del and a bunch of other ferry pilots were grounded by the weather in Dallas. It had been a long leg, and they were ravenously hungry. As soon as they were checked into the hotel, they converged on its dining room *en masse*. Del was in the back of the group, and a waiter had already begun seating some of the men when she was stopped by the maitre d', who said, "We do not serve women wearing pants." Del was the only woman in the group, so she thought the decent thing to do would be to strike out on her own and let the rest of them dine in peace. But before she reached the lobby, she heard the sound of chairs being pushed back from their tables. She stopped and turned around. The men who had already been seated were standing. Every single one of them was following her out of the restaurant and out of the hotel. As they walked along the sidewalk one said simply: "You're an Air Force pilot serving your country." And because Pilot Scharr could not enter a hotel dining room in trousers, the entire group choked down a meal at a diner down the street.

Meanwhile, at Howard Hughes Field outside Houston, it was becoming pretty clear to everybody that a program as big as the Woofted couldn't be kept secret. In fact, the basketball team story had worn pretty thin—most of Houston was aware of what was going on. Cochran decided to introduce the twenty-three graduates of the first class to America at their graduation. The other two classes, a total of more than one hundred trainees, would march in the graduation parade.

But the women still had no official uniform, and so decided to order one through the Ellington PX—white open-collar shirts, khaki pants, and matching khaki soda jerk hats they called "overseas caps." It was Mrs. Deaton who realized that the twenty-three would have no wings. Here they were, pilots about to become graduates of a flight school, and nobody had thought to remember *wings*. So Mrs. Deaton looked for Jackie, but she was away somewhere again, and not knowing what else to do, she phoned Floyd

Odlum. Floyd didn't miss a beat. Practically before she could get the question out he said, "Send me the bill." Mrs. Deaton and Lieutenant Fleishman purchased twenty-three standard issue AAF wings at the PX, took them to a local jeweler and asked him to replace the center shield with the Woofted class and training detachment number.

On April 28, 1943, Ellington Field looked different—there were flags and Army brass in a reviewing stand and, strangest of all, *reporters*. They were everywhere. Reporters from local papers, reporters from the United Press, reporters from the Associated Press, even cameramen from Movietone newsreels. The classes marched, the generals made speeches, and the wings were awarded. Through it all, Jackie Cochran stood there like the proudest mother in the whole world, while the cameras clicked and the flashbulbs popped.

Jackie Cochran had another reason to be pleased. Recently, she had received notice that the Army Air Force needed 750 pilots by the end of the year, and another 1,000 the year following. Toward this end the WFTD would be given its own training facility at Avenger Field in Sweetwater, Texas, and all the PT-19s and AT-6s it could handle. It seemed clear that despite the reservations that the Army Air Force had had when all this began, the Woofteds were now legitimate.

Avenger Field was small, as airbases went. It had two criss-crossing, three-thousand-foot gravel runways, and a few low, wood-frame buildings. These were administrative offices, officers' quarters, a ready room, classrooms, and a taller building for testing parachutes. A control tower and a new hangar were under construction. There were also two long buildings built side by side and divided into rooms called "bays." These were trainees' barracks.

Training itself at Avenger was much like that at Houston, though learning to fly on instruments was slightly different. Here they used simple flight simulators named Link Trainers. Seated beneath a hood which prevented her from looking outside, a trainee followed instructions she received over earphones. Using only the instruments, she manipulated the controls which caused the machine to rotate on a pivot. When a student had performed well enough on the Link, her trust in instruments was insured by a similar exercise in an actual airplane. An instructor rode in the front seat, ready to take the controls, while a black hood was pulled over the student's section of the cockpit, so she could see nothing but the instruments—clock, airspeed indicator, altimeter, and compass. The exercise was exacting: She was expected to fly a precise four-leg pattern, each leg the same length. The instructor would ask her to climb to 500 feet on the first leg, make a turn of exactly 450 degrees, fly level on the second, make a turn of 270 degrees, descend 500 feet on the third, make another turn of 450 degrees, and then fly level to the point of origin. Trainees practiced stalls under the hood, as well, an exercise which required them to rely solely upon the turn-and-bank indicator needle and the airspeed gauge for recovery.

In the spring of 1943, at any given time, there were three classes in various stages of training, for a total of perhaps four hundred trainees altogether. During daylight hours there were as many as fifty planes in the local airspace at any given moment.

At Avenger Field, like a lot of military bases, luxuries like butter were in short supply. At the cafeteria, in fact, there was a sign that read Butter Is Critical. To the girls this seemed pretty funny, and so some of them, sliding their trays along the cafeteria line, started saying, "I'd like some critical butter, please." Still, these women had been children through the Depression, and they had learned to make do or do without, so they were loath to take any good fortune for granted. Whenever one student started to complain, another would say, "Kid, we never had it so good." The first might laugh a little at herself and then give the standard return, "Yeah, a roof over our heads, three squares a day, and a loaf of pumpernickel."

In April the thermometer in west Texas reached one hundred degrees, and it would stay there until September. The barracks did not offer cross-ventilation, and at night the air could be suffocating. A few of the girls asked Mrs. Deaton for permission to put their bunks outside. She allowed it, and all the girls did it, lining the bunks up in military rows in the open ground between the barracks. Early one morning one girl was making her way to the latrine when she stepped on a rattlesnake. When she realized what it was, and where she was, she let out a scream that might have curdled milk. The little rattlesnake, no doubt as terrified as the girl, got away—but not before it had permanently changed sleeping arrangements at Avenger Field. Now there was a sort of Hobson's choice. A trainee could stay inside the barracks, and spend the night on sweat-soaked sheets. Or she could have fresh air, and take her chances with the local fauna. It was a little cooler outside because there was a breeze, but she had to get under the blanket and then tuck it in all around.

There was an official Woofted motto at Avenger Field: "We live in the wind and sand . . . and our eyes are on the stars." It wasn't long before the girls started adding ". . . that is, when the wind and sand aren't in them." In west Texas, the wind could pick up suddenly. The light trainers could actually be lifted by sudden gales, and sometimes the commanding officer would have to order all the girls out on the field to put all their weight on the wings, two to a plane. On other days the sky would be black with grasshoppers—whole clouds of them, right out of an Old Testament plague.

The military grapevine being what it was, there was plenty of talk about the first all-female flight school in history. Aircraft from nearby training facilities were developing all sorts of problems and having to make "forced" landings. On one afternoon alone there were thirty-nine. The flight controller on the airstrip at Avenger would hear from flights routed nowhere near Sweetwater. Often it would be some cadet with a vague description about a possible problem with his aircraft. He would land, and roll on up to operations. While his plane was being checked out maybe he'd walk over to the PX for a Coke and a cigarette. When he would get back to his plane the mechanic would tell him they couldn't find a thing wrong with his engine. He'd look at the mechanic and look at his plane and scratch the back of his neck and maybe grin a little and say, "Guess I musta been hearin' things."

This went on for months before Cochran decided that except in the case of actual emergencies, no planes from other bases were to land at Avenger Field. This isn't to say that the girls didn't still have fun. They began to call it "Cochran's Convent," and put up a sign near the showers that read "Men Trainees Only" and had their pictures taken standing near the sign.

The training went on. The girls were expected to practice marching drills. To most of them, marching seemed quaint and pointless. In fact, the girls' sentiments on this subject were well in line with those shared by military pilots everywhere—specifically, that marching was fine for infantry. But a pilot didn't have much to do with infantry, except for those days, with the wind and sun in her face and the clear cold air in her lungs, when she banked her plane a little and looked over the side so she could see them crawling like so many ants up the side of a mountain, the same mountain that was already a long ways behind her.

Still, they were expected to march. Some of them thought that, like the infantry, they ought to have songs. So the Woofteds began to sing. A lot of the stuff sounded like a female version of songs from small-town Fourth of July parades. "We are Yankee Doodle Pilots, born with a yearning to fly," et cetera. Some of the songs, though, were more strident, and some were downright ribald. Jackie was afraid—maybe for good reason—that such high-spirited antics might compromise the program. It was, after all, 1943, and these women—many of whom had never lived away from their parents—were living on a base, and there were men on the base too. And among the taxpaying public there were mutterings that some of these women were "Airport Annies." The last thing Jackie needed was a scandal.

The girls themselves had more immediate concerns—like who was going to graduate and who was going to wash out. It was this particular concern that drew a great deal of attention to an architectural feature that made Avenger Field different from most American military bases in 1943—the Wishing Well. It was circular, about twenty feet across the middle. Its wall was stone, about three feet high, and the water came up almost to the edge. One day somebody tossed a coin in it before a check ride or before a test, and it seemed like such a good idea that soon everybody was doing it. It became a small sacrifice to whatever fates determined the unknowns of instructors' tempers and fickle west Texas weather and unresponsive controls. It wasn't long before another tradition developed—throwing a trainee in after her first solo flight. This became the unofficial Avenger Field rite of passage, and every trainee wanted to be thrown in because then she knew for sure that she wouldn't wash out of the program.

The girls at Avenger had heard about ferrying graduates and WAFS, and some had made friends in earlier classes, now graduated and stationed at Long Beach or Romulus. But few of them had actually seen a woman ferrying pilot. One day some of the girls were in the airport, empty except for a smallish figure sleeping on a bench. As they neared her, their voices became hushed. One of them whispered, "It's a WAF." Wearing her wrinkled, oil-stained flight suit as pajamas, she was asleep on her packed parachute. They moved past her silently and reverently.

13

FERRYING

In fact, by this time Nancy Love's WAFS pilots were criss-crossing the country ferrying planes, and falling asleep where and when they could. And sometimes ferrying wasn't merely grueling and exhausting. Sometimes it was also frustrating. If a group of WAFS had orders to make a delivery on contact flight, and were socked in, sometimes they'd wait for days.*

Ferry pilots were required to report delays to their home base, but because movements of aircraft were secret, telegrams were coded: "THREE," for instance, meant "grounded by weather." In the first few weeks of ferrying, Betty Gillies and B. J. Erickson were grounded by a storm in Albany, Georgia. Betty made a practice of sending a second telegram to her husband, Bud, at his office at Grumman. Betty had passed four long wet days in Georgia, and Bud had received telegrams with the message "THREE" for three consecutive days. On the fourth day she heard the same forecast. That day, Bud received a message from Betty no one understood.

*The small trainers like BT-13s lacked navigational instruments, and required a pilot to navigate with only a compass and map. It was called a "contact flight" because she sighted or "made contact" with landmarks along the route.

It read "HEBREWS THIRTEEN EIGHT." In the Second World War the Allies used many codes—they could get pretty elaborate. Navaho Indians trained as communications officers were using the Navaho language itself. But Bud Gillics and his friends were aircraft designers, not cryptographers—so they looked at this piece of paper, they turned it upside down, they got pencils and made anagrams. Suddenly one of them went away and came back with a King James Bible. He turned to Hebrews 13:8, and showed it to Bud, who smiled. It read "Jesus Christ the same yesterday, and today, and for ever."

On days when the WAFS were grounded, they grumbled and joked, and found out things about each other they'd never otherwise know. Barbara Poole was an intimidating poker player. Evelyn Sharp could walk on her hands. She would cross a room, talking to you all the way, her legs straight up in the air. Then she would turn around and do it again.

By early spring, word of Nancy Love's checking-out in the P-51 Mustang had spread throughout the WAFS. At Dallas, at Long Beach, and back at New Castle, women were asking commanding officers to be allowed to make transition. Betty Gillies had transitioned to a P-47 Thunderbolt. The plane's cockpit was sized for longer legs, so she had tied wooden blocks to the rudder pedals. Before long Gertrude Meserve and Teresa James were given check flights in the P-47. And there was no slowing Nancy herself. By early May, she and B. J. Erickson had transitioned to the C-47 twin-engine cargo transport. Soon transition was all the talk among the WAFS and the WFTDs. At Avenger Field, cadets not even through primary training would lie awake on their bunks and debate the relative merits of flying bombers or pursuits.

And at Romulus Air Force Base, squadron leader Del Scharr had her eye on a particular aircraft a lot of pilots wanted nothing to do with.

The Bell P-39 Airacobra had a 20mm cannon in the nose, and the end of its barrel came straight out of the propeller hub. To allow room for that cannon, the engine was placed just aft of the cockpit.

This single design variation was the cause of some big problems. For one, the weight was balanced differently. Some pilots were afraid that if they nosed over, the engine would crush them. There were other differences too. It was armor-plated, and that made it heavier and difficult to maneuver. Simply to stay airborne the plane had to maintain a relatively high speed. There had been crashes, several fatal. Airacobra ferrying pilots had dropped right off the bottom of the actuary tables. It wasn't long before the men were doing all they could to avoid those assignments.

Still, pilots were getting checked out in the P-39, and pilots were ferrying it. In fact, there was a certain bravado about it, and at Romulus there were jokes. One pilot would make his elbows stick out behind him, waddle in a circle like a duck and say "Are you noivice in the soivice? Are you noivice in the soivice?" The boys at the officer's club would just crack up at this. They even had a song:

Don't give me a P-39
With an engine that's mounted behind
It will tumble and roll
And dig a big hole
Don't give me a P-39.

Nobody knew exactly who started it, but regarding the P-39 there began one of the instant traditions that are common to the military. A pilot who had checked out in a P-39 was accorded the privilege of flying with a red baseball cap. It was a modest emblem for bravery, but then bravery was something pilots were modest about.

Like anyone at Romulus, Del heard a lot of talk about the dangers of the P-39, some of it pretty direct. One civilian pilot, wearing a bandage on his foot, had bailed out of an Airacobra somewhere over the Appalachians. He was behind her in the chow line at mess one day, and he asked her to sit with him. Del located an empty table and carried her tray toward it, and he limped over behind her, balancing his own tray on one arm. When he got settled, he began telling her about the accident. He got very melodramatic about it all. As his eyes welled up with tears, he told her, *Don't fly that thing, think of your husband.* But the tall woman sitting across from him was utterly unmoved. Del knew that most accidents happened through carelessness. She had long believed that if she died in an airplane, it would not be because she was its pilot. And now she thought to herself, so what if five P-39s flying in formation had crashed? *They followed their leader into a cloud!* It was not only against Ferrying Division regulations and the CAA—it was against common sense! If this guy had performed a proper preflight, his engine wouldn't have cooked itself, the AAF would still have a plane, and he would be doing the jitterbug at the Saturday night base dance. But he was still talking, and she looked at him with the clear eyes of homesteader women in those old sepia photographs, the eyes that seem to say "Mister, I'll believe none of what you tell me and exactly half of what you show me." Finally, she just interrupted him, saying "Did you check your coolant before takeoff?"

He, like the fourth grader admitting to the teacher that he had not washed his hands, replied, "Uh, no."

In 1942 many people had fairly abstract notions of what it took to be a pilot—courage, luck, natural ability, or some combination thereof. But in her core, Del Scharr was a Missouri schoolteacher. As far as Del was concerned, there was no problem in aviation or anywhere else that couldn't be solved by a little bit of hard work. Del tackled her assignment to the P-39 the way a conscientious high school student might tackle a particularly stubborn problem in long division. She knew that the worst problem posed by the P-39 design was that the engine behind the cockpit made it harder to cool, especially at slow speeds. Once a plane was flying, air rushing into the airscoop and out the exhaust carried most of the heat away. But P-39s were overheating sitting right on the runway. It occurred to Del that the preflight might be rearranged in such a way as to allow the pilot to get airborne before the engine became dangerously hot. To do this, she would have to understand the plane. Del asked the major in charge of the school for material on the P-39. He said that the plane was too new—there was no material yet. But Del knew that there are more ways into the woods than one, and she paid a visit to Bell's company representative on the other side of the field, asking for everything he had on the P-39.

In the 1930s and 1940s aircraft engines were generally unreliable, a situation which demanded that pilots also be competent mechanics. Del knew all about cotter keys and types of screwdrivers. She had changed oil on a Kinner 125-horsepower engine, and she knew how to feel for babbitt as oil leaked out of the crankcase. She could take off the rocker arm boxes, test the filters for warp, check the valves, and set the valve clearances. In short, Del Scharr was no stranger to the powerplant of the modern airplane. But as she sat cross-legged on her bunk and looked at the papers in front of her, she began to realize the P-39 Airacobra was something else again. It was more complex, more sophisticated than anything she'd known. Still, Del had entered new areas of study before, and she could do it again. So she began at the beginning—in fact, she began where the plane began, with its constant speed propeller. Even this was difficult. Del had never seen the inside of a complicated airscrew, and the little booklet on the propeller might as well have been in Sanskrit. There were cutaways, exploded diagrams, and current flow charts. She read the same paragraphs again and again, then studied the drawings and matched part *A* from the exploded view to section *A* in the text. Slowly, she began to understand. She sat cross-legged on her bunk with a legal pad and a pencil and took the preflight checklist . . . and she started rearranging it. A few of the necessary changes were obvious—all inspection had to be done before the engine started—and rather than check the flaps as she waited for takeoff she could do it while she taxied. That would save a few minutes. Del sat and chewed the end of the pencil and crossed things out and crumpled pages and started over. After a few hours the pencil was chewed to a nub, and the floor around her bed was littered with crumpled yellow paper. She was tired, but she was also a little elated, because she had in her hand a preflight procedure that should allow the P-39 to get airborne before its engine overheated.

On the afternoon of June 28, she walked out onto the flight line where her P-39 was waiting. She did a careful walkaround, then climbed into the cockpit. The layout of the instrument panels was standard. On the left were trim tab controls, landing gear, flaps, throttle, mixture and propeller pitch levers. Straight in front were flight and engine instruments, and fuel and engine switches. And to the right were circuit breakers and the radio. She set the mixture control to "IDLE CUT-OFF," pushed the propeller control forward, opened the throttle, turned on the battery switch, and then looked around to make sure the area around the propeller was clear. She set the fuel booster pump to the "On" position. She engaged the starter. The propeller began to move slowly. She turned the ignition switch to "On." Immediately the propeller blurred into motion, there was a deafening clatter, and the exhaust stacks barked. She brought the fuel mixture up to "Normal," set the throttle at 1,300 rpm, and let the engine settle into a steady thunder. Outside, the line crew was standing by. She gave them the thumbs-up, and they removed the wheel chocks. Now she was moving, steering with the tail wheel. She finished the checklist. The warning lights were functioning, the flaps were up, and the parking brake was off.

Soon she was climbing. Through the forward section of the canopy she could see only clear sky. With her left hand she reached for the landing gear lever, pulling it inward and up, and she felt the double thump of landing gear retracting. As her speed rose she felt a gentle push into the back of the seat. She was at 175 miles per hour, still gaining speed, and still feeling her back pressing into the seat. She glanced at the engine temperature and pressure gauges, set the oil and coolant doors to "Automatic," and turned the fuel booster pump to "Normal."

She leveled off at 10,000 feet. The sky was summer blue. A few wispy cirrus clouds floated above her. The Airacobra was responsive, but sure. She tried a stall, pulling the throttle back until the engine quieted a little, and then pulling back on the stick. The nose began to pitch up, there was a little buffeting and the plane began to roll left. The propeller still wanted to torque the plane. She released the stick, the nose dropped again, and the plane rolled back to level flight. Recovery from stalls, she noted, was immediate. Del made a few circuits of the field, allowed herself a little smile, and prepared to take it in.

Later that afternoon Del was in the PX, trying on red baseball caps in a mirror. She was pulling the bill lower over her eyes and tucking in a stray strand of her hair, when she saw one of the male trainees standing behind her. He said to her reflection, "You're not really going to *ferry* them, are you?" Del, a little caught off guard by this, answered without turning around and almost without thinking, "What did you think that was—a publicity stunt?" Before the young man could think of a response, she was at the cashier, saying, "I'll take the hat."

The Lend-Lease Act of 1941, which allowed the Allied nations to lease U.S. equipment for the duration of the war, contained provisions for the Soviet Union to lease a number of pursuit aircraft, among them the P-39 Airacobra. WAFS and other ferrying pilots flew P-39s from the Bell plant in Detroit across the border to Canada, where they were picked up by Soviet pilots and flown west across Canada, then Alaska, then over the Bering Sea to a base in the Soviet Union.

The Soviet Union had women pilots in actual combat—not just a few, but three whole regiments. The 46th Night Bomber Regiment was particularly proud of the fact that it was entirely female. Everyone from the commander to the electrician was a woman. They called themselves the "Night Witches," and were engaged in the fiercest aerial combat in the war. The tactics themselves were dangerous. They would fly without lights in the dead of night, and as they neared the target they would cut their engines, drop bombs during a glide, then restart the engines. As might be expected, casualties among pilots were common. The most famous of these pilots was named Lilya Lityvak; she was a flight commander with twelve personal kills and three shared kills. Lityvak was made Hero of the Soviet Union, a title awarded almost automatically to a pilot who had flown five hundred successful combat missions. In any other regiment, a single award to a single pilot would have been cause for pride. By 1944 the Night Witches had earned twenty-three. Although the women were written about in newspapers, and a few reports leaked to the West, some thought the stories of Russian women pilots were propaganda. The WAFS who ferried P-39s across the border knew better.

On the flightline at a Canadian airbase, a tired WAFS pilot saw another woman, a nineteen- or twenty-year-old Russian pilot wearing jodhpurs and a flying tunic. In the pale, clear morning light, with the smell of raw gasoline and breath visible in the air, the WAF watched the Russian woman climb into the cockpit of a P-39. In a few minutes, when the plane was cleared for takeoff, it roared down the runway, slowly rose, and then disappeared against the western sky.

By early July the buses unloading 30 or 40 young women at an Army airbase in the middle of west Texas had become almost routine. As usual, the newcomers arrived with more than could be fit into a trunk; as usual, they were surprised at the primitive surroundings; and as usual, they were likely to regard members of the earlier classes with great respect. Soon they learned that the class of 43–4 counted among its 151 members a young woman remarkable even by the remarkable standards of twenty-year-old women pilots. She was Chinese American, and her name was Ah Ying Lee. Her father had settled in Portland, Oregon, married, and begun a family. Ah Ying and an older brother had begun to fly in 1932. In 1937 Japanese troops invaded from Manchuria, and Ah Ying, her brother, and nine other Chinese-American pilots boarded a ship to China, intending to join the Chinese Air Force. The Chinese accepted the men, but they would not train a woman pilot. For a while Ah Ying taught school in the village where her father had been born, but it was difficult. She had grown used to American freedoms. So in a few months she returned to the United States, this time to New York. And in the fall of 1942 she heard that Jacqueline Cochran was training women to fly for the U.S. Army Air Force.

By early summer Ah Ying was well into advanced training. One day she was flying an AT-6 over the west Texas countryside when the engine quit. She managed a dead-stick landing in a field. As she climbed out on the wing she saw a man running toward her. He had a pitchfork, and he was turning his head back over his shoulder yelling at whoever might be in earshot, "The Japs have landed, the Japs have landed!" She screamed back, "Me no Jap! Me yankee! Girl pilot!" And he stood in front of the wing, breathing hard, those three parallel tines at the ready. "No you ain't, don't you move!" She got angry, shouting, "Me damn Yankee girl! *Me damn Yankee girl pilot!*" In time the man was joined by others, hired help and neighbors. Ah Ying screamed, "Call Avenger. Call Avenger Field!" They had heard of Avenger, and one went and made the call. Half an hour later a staff car drove up. An officer got out and said to the small crowd around the plane, "See the Air Force insignia on her sleeve? She's an American pilot." But they could not understand how this man could be so naive as to believe an insignia could not be copied by the enemy. The officer used his radio to call the base again. More officers arrived. Finally Ah Ying was surrendered like a prisoner of war.

By the summer of 1943 hundreds of graduates of the first few Woofted classes had joined the WAFS squadrons around the country. They lacked a name. To call them Woofteds was inaccurate insofar as they were no longer in training. And to call them WAFS would soon become inaccurate, because plans were in the works for other types of work, and a WAFS was a ferry pilot. The Air Force needed a name for all its women pilots, regardless of assignment. In Washington, Cochran and Arnold discussed alternatives, and decided that Women's Airforce Service Pilots described the increasingly broad roles Jackie had envisioned, and yielded an acronym with a satisfying military snap. On August 4, 1943, copies of AAF Regulation 20–8 were distributed to all commands, announcing, "The title WASP is hereby designated for the women pilots of the Army Air Forces." Despite the inherently plural nature of the term, it was also used as a singular noun to describe a member of the group.

Cochran, based in Washington, had on July 5 been named Director of Women Pilots within the AAF. She would be special assistant to Major General Barney Giles, assistant chief of staff for Operations, and she would be based at the Pentagon. Meanwhile, Nancy Love was transferred to Ferrying Division Headquarters in Cincinnati, and made WASP executive on the staff of the Ferrying Division, a position that give her jurisdiction over all women ferrying pilots. She remained under the direct command of Colonel Tunner, head of domestic ferrying. Technically, Love was also an indirect subordinate of Cochran, but their rather different responsibilities would give the two women little cause to communicate in the months ahead.*

On July 19, 1943, in a conference room at the Pentagon, twenty-five recent graduates of Avenger Field watched Jackie Cochran, standing at the head of the table. "You are going to be assigned to a top-secret mission," she said, and paused. "You are going to have a chance to fly bigger and better airplanes than women have ever flown before." The twenty-five women seated around the table waited, expecting more. But Cochran simply asked them a question: "Is there anyone who doesn't want to be assigned to this mission?" Military pilots were familiar with this—the tone, the atmosphere, the question itself. It was a request for volunteers for especially dangerous combat. The implication, also familiar to military pilots, was clear: Nothing would be held against you should you not volunteer, and nothing would appear on your record. Cochran looked slowly around the table and her gaze met the eyes of each of those twenty-five women in turn. A few moments later it was over. No one had raised a hand.

They would be stationed at an airbase on the Atlantic Coast near North Carolina's Cape Fear. It was called Camp Davis, and it was as big as a medium-sized city—fifty thousand officers and enlisted men were there to learn to operate antiaircraft artillery. The WASP were to fly planes towing targets for that artillery, large muslin or silk sleeves, maybe thirty feet long, on the end of fifteen-hundred feet of steel cable. As this was the last training the gunners would have and because there was no way to know whether they were hitting the target except to count the holes afterward, they used live ammunition, bullets dipped in colored wax so they could identify who made the hits.

Every antiaircraft gunnery crew seemed to have a joker who would say over the radio to the tow pilot, "I see the target—but what's that thing *behind* it?" In 1943 anybody towing targets for the Army Air Force had heard that one a hundred times. It was as dangerous as it sounded. The big guns were heavy and awkward.

*Because the first WAFS to arrive at New Castle shared a certain pride in their status, they were allowed to retain their original designation and uniform.

Sometimes the guns and the gunnery crews were in vehicles. They would go bouncing over the North Carolina back country and one tire would hit an exposed root and the vehicle would bump a little and a shot would go in the wrong direction, and then the vehicle would lurch into a ditch and another shot would go in another wrong direction. A few planes came back with bullet holes in their tails, and every now and then a tow plane would have to make an emergency landing because, for instance, an engine was shot out or a pilot caught a bullet in the foot. One WASP, returning from flying a tow-target pattern, found bullets in her plane, and asked the mechanic to give them to her when he dug them out. He did, and she took them to a jeweler and had them made into a bracelet. The other girls loved it. There was something of the warrior about it—like making the teeth of an enemy you killed into a necklace.

But at Camp Davis, stray ordnance was only one of many dangers. A great number of planes were red-lined. There were bad radios, flaps that responded slowly, and on one plane the canopy stuck. But none of this was deemed serious. The overworked mechanics, short on parts, gave their energies to maintaining the engines, thinking that so long as the engine ran, the plane would stay in the air, and the other problems would be, at worst, annoyances.

Gunnery crews at Camp Davis were trained to operate at night with searchlights. Planes flying tow-target were expected to fly at night too. Training at Sweetwater had included plenty of night flying. They used to outline the landing strip with oil burners—it was hard to miss. But like everything else at Camp Davis, night flights were harder. It was an East Coast base, so it was blacked out. They didn't even light the landing strips. In daylight, a pilot at 3,000 feet who needed to know about crosswinds on the ground just looked at the windsock near the tower or the tops of the sea pines. But at night a pilot couldn't see either very well, so she had to rely on the altimeter, though that gave only a rough estimate. To make a landing she had to come in low—maybe twenty or thirty feet off the ground—if she expected to see the end of the airstrip. The airstrip, however, was surrounded by pines, and some of them were seventy feet tall.

On a warm evening in late August, Marion Hanrahan was flying tow-target in an A-24 over Wrightsville Beach. Ahead of her was a dark silhouette barely visible against the night sky, a little above the darker line of trees. She knew it was a second A-24 piloted by Mabel Rawlinson, a former librarian from Kalamazoo. Then suddenly something wasn't right . . . the other plane seemed to stop in midair . . . and Marion couldn't see the shape at all. It was there and then it *wasn't* there. All she could see was the line of trees darker than the sky and then over the sound of her own engine she heard the sound of another engine descending too fast. An eternity later there was the sickening sound of wrenching metal, and suddenly everything was illuminated in a nightmarish orange light—the sea pines, the airfield, reflections from the water. On the ground below her and ahead of her she could see the patch of bright flames. Marion was flying right over it and she could see the plane itself—it was broken in two. She could see it beneath her. And she could hear a muffled scream.

Marion brought her plane in. On the ground, in the barracks, they had heard the sputtering engine and had run outside. There were sirens, crash crews. It wasn't long before everybody knew. The instructor had been thrown clear, but Mabel had not been so lucky. She had been trapped inside the cockpit while the plane burned. It was the one with the canopy that stuck. And the muffled scream that Marion had heard was from a young woman in agonizing pain trying desperately to force open a canopy while smoke filled the cockpit, a woman who in a few moments would be burned beyond recognition.

The next day Cochran and her assistant Ethel Sheehy flew to Camp Davis. The WASP were told they would meet with Cochran at 1900—seven that evening. In the afternoon, Cochran would meet with the officers of Camp Davis.

Meanwhile, the WASP talked it through. Some accepted this as the price of a nation at war. It was bound to happen sooner or later. Others were, as they say, fit to be tied. In particular, Marion, Kay Menges, Mary Lee Logan Leatherbee, and Marie Shale were angry—and they had choice words for Cochran or anybody else who was about to tell them this place was run by the book. That long afternoon they paced the quarters. The assignment at Camp Davis had been presented to them as a top-secret mission for elite flyers, but you didn't have to be a career officer to see that it was being run in a way that was inept and slipshod. One of their own was now a charred carcass in a coffin because a canopy latch had not been repaired. And tonight they were in no mood to put up with any bureaucratic . . . guff. "Slice it as thin as you want," they would say, "it's still baloney."

At fifteen minutes before seven, the twenty-four WASP waited in a room in ops. Some of them sat in the folding chairs and stared straight ahead. A few of them paced and smoked. They had been talking all afternoon. To each other, they had nothing left to say, but for Jackie Cochran, they had words.

Cochran didn't know what was going to happen, but she was prepared for the worst. Her best pilots, their number diminished by one, were quite capable of ending the entire WASP program. It would take only a letter or two to a congressman, a call to a newspaper. So she was prepared to hear them out. At 1900 hours she walked through the door. Sheehy was with her.

What followed was what diplomats call a frank exchange of views—or more accurately, an expression of one view, because Jackie Cochran mostly listened. What her WASP told her about conditions at Camp Davis was sobering. Engines were quitting in midair—it had happened twice on the day of the accident. Since the WASP had been at Davis eleven pilots had made forced landings, and every pilot at the base had become used to searching for pilots downed in the swamps. The mechanics were overworked, one of them had said that at most three—*three*—planes on the whole base were airworthy. But it wasn't just the machines. The instructors didn't seem like proper instructors—the WASP had more flight time than most of them—and it was well known that quite a few of the instructors were assigned to tow-target because they had washed out of more prestigious programs. At this last, Cochran had to ask, "Who agrees?" There were a lot of hands in the air. Cochran nodded, and promised an investigation after the ceremony for Mabel.

The next day there was a second meeting, with Cochran and Sheehy, but also with the base commander, the chief surgeon, and the public relations director. The women were quiet as the commanding officer rose and walked behind the podium. He was smiling. He cleared his throat. He said he was delighted to have the WASP at Camp Davis. Then he made a little speech about how if there was anything he could do to make them comfortable, they need only ask, and all the while he smiled this quietly insane smile. When he was done the chief surgeon rose to speak, and he was wearing the same quietly insane smile. He had his own little speech, and he announced that the WASP would be allowed to use the nurses' quarters.

Not a word about Mabel. In one of the back rows Marion looked at the friend to her left and whispered, "This is nuts," and her friend, who did not turn her head, whispered back, "I know, but that's the commanding officer . . . and we're in no position to tell him how to run . . ." Before she could finish Marion was standing up. The twenty-three women turned to look. Cochran and Sheehy, at the front of the room, were looking too. The droning speech from the flight surgeon at the front of the room stopped. The surgeon stopped smiling. He looked out over the heads at Marion and said, "Yes?" She swallowed hard and said, "That's not enough." It didn't seem to matter. Cochran assured the women that the planes were airworthy and made some vague promises that matters would improve. Then she and Sheehy left.

And so Marion wrote Cochran a letter of resignation. Cochran would not accept it—for Jackie, it might be the beginning of the end of WASP in assignments beyond ferrying. So Marion, now joined by Marie Shale, requested that the Ferrying Division, from which they had been released for assignment to Camp Davis, accept their requests for transfer. To Cochran, this was nearly as bad, and she blocked the move. After a few weeks Cochran arranged for both resignations.

Ferrying was for the most part a pretty straightforward business. A pilot picked up a plane, flew it to its destination, and then returned to base. The time this process took could vary considerably. A pilot on the New Castle-to-Newark run might be able to make several deliveries in a single day, but a round-trip from Long Beach to Atlanta could take more than a week.

In the summer of 1943, Teresa James was stationed at New Castle, and one morning she received orders to deliver a P-47 to Evansville, Indiana. The P-47 was fast enough to make this trip in just a few hours, and Teresa was told that she could hitch a ride back to New Castle on an Army Transport plane that same afternoon. So she took off in her WAFS-uniform shirt and slacks with nothing more than a tube of lipstick in her pocket. When she arrived in Evansville, she found that she would not return to New Castle quite as soon as she thought. Evansville had another P-47 that needed to go to California the next day, and wanted pilot James to take it there. While mechanics tinkered with the plane's hydraulic system, the next day stretched into five days, and Teresa was stuck in Evansville with nothing to wear but the clothes she had arrived in. The shirt was wrinkled and the trousers were getting stretched around the seat and the knees. At last the P-47 was deemed airworthy, and two days later Teresa arrived at Long Beach—where they asked her to take a P-51 to Fort Meyers, Florida. Off she went, headed across Texas and Louisiana and Alabama, where, in summer, thunderstorms roll in just about every afternoon, and so she could only get so far every day before she was forced down by the weather. Every morning she put on the same shirt and the same trousers. She tried to wash them out each night, but in the humidity they never dried well. Because she hadn't brought her uniform jacket she was not allowed to dine in officers' clubs, so she subsisted on sandwiches in mess canteens. And because she was looking rather disheveled and because she had no luggage, hotel clerks hesitated about giving her a room for the night. When she finally got to Florida, she thought, surely they'll take one look at me and send me back to New Castle at last. Instead, she was handed orders to deliver an AT-6 to a base in Oklahoma, and in Oklahoma they gave her a P-39 and told her to take it to Great Falls, Montana, and in Montana they walked her out to the line and pointed out a P-47, and Teresa thought she was going to spend the rest of the war in the same pair of trousers. When they told her the P-47 was needed in New Castle she thought she was hearing things. Thirty days, seventeen states, six airplanes, and eleven thousand miles after she had left New Castle for an afternoon's journey, Teresa James finally made it back. She walked into BOQ 14, and the women she met in the hallway stared at her in astonishment. "Jamsey," one of them said, "You look . . . you look like the wreck of the *Hesperus*!"

On July 1, 1943, by an act of Congress, the Women Army Auxiliary Corps was redesignated the Women's Army Corps (WAC), and granted full military status. Many assumed that when the WAC was militarized, the WASP would be brought into it. But Jackie Cochran believed that such an organizational shift would entangle her pilots in Army bureaucracies, and hinder her plans to broaden their duties. It would also make her a subordinate of Oveta Culp Hobby, the director of the WAC, and a woman with whom Jackie had personal and professional differences. Jackie told Arnold it might be better to wait. Arnold had, up to this point, been undecided. If the women could achieve military status and its associated privileges only through the WAC, then perhaps the merger was worth it. His hesitation lay in a belief shared by most in the Air Force, that pilots were fundamentally different than members of other services, if not actually superior to them. And if the director of Women Pilots believed that assimilation by the WAC was unsatisfactory, that clinched it. Arnold agreed that if and when the WASP were militarized, they would become part of the Air Force.

In the summer of 1943, Representative John Costello of California began drawing up a bill that would give the WASP the privileges, insurance and death benefits accorded male officers and cadets in the Army Air Force. A militarization of the WASP seemed imminent.

14

THE TROJAN WOMEN

In the summer of 1942 the Air Transport Command had begun Operation Bolero, the mission of which was to ferry P-38s, B-17s, and Douglas C-47s across the Atlantic. The route was a series of hops—Presque Isle, Maine, to Goose Bay, Labrador, to Greenland, then Iceland, then on to Prestwick, Scotland. Officials expected losses of ten percent of planes dispatched. Actual losses would be only half that, but it was never easy. The second delivery, six P-38s and two B-17s, ran into bad weather, and was forced back before it reached the east coast of Iceland. The intended landing site in Greenland was closed. The pilots were disoriented by the storm and low on fuel, so they so sent out an SOS indicating their position and decided to put down on the Greenland ice cap. One P-38 pilot, thinking the ice looked hard and smooth, came in with landing gear lowered, but flipped his plane. So the rest of them came in wheels up, and simply skidded onto the ice. Then all the twenty-five men had to do was get in the two B-17s, try to stay warm, and wait. Nine days after the landing, a rescue team arrived by dogsled, and the twenty-five abandoned the planes. With their rescuers, they marched southeast to the coast, where they were met by a Coast Guard cutter. No lives had been lost, but it was a reminder that even in the air age, the North Atlantic still had teeth.

In summer of 1943 the Eighth Air Force in England requested one hundred B-17s for forays deeper into Europe. Colonel Tunner had plenty of pilots newly qualified in the B-17, but they were—there's no delicate way to put it—afraid to jump the pond. Tunner had long since put aside his doubts about women transitioning to pursuits and bombers. Of late, he had grown more and more impressed with the WAFS. A few weeks ago, Del Scharr's solo in the P-39 had encouraged male pilots to fly a plane which they had been avoiding, and he thought the trick might work again. Because Hap Arnold was in England, Tunner cleared the plan with AAF chief of air staff Barney Giles. Then Tunner called Nancy Love in Cincinnati and said, "I want you and another WAFS to jump the pond." The flight would be secret. If something did go wrong they wouldn't need the newspapers reporting it, and if all went well, the grapevine among ferrying pilots was efficient enough that the good news didn't need the papers. And so, for the second time in eight months, Nancy Love phoned Betty Gillies about an opportunity to make some aviation history. Betty got the wood blocks that she would tie to the rudder pedals so her feet could reach them, and the two WAFS prepared to pilot and co-pilot a B-17 across the North Atlantic.

Because three B-17 deliveries were necessary to qualify a pilot for an overseas delivery, Nancy and Betty spent a few weeks that summer delivering Flying Fortresses to bases in the Midwest. And then, on September 1, Flying Fortress E-30624, piloted by Love and Gillies, and carrying a navigator and flight engineer, rumbled along the runway and into the air, bound for Presque Isle. They made good time, but bad weather grounded them in northern Maine. On September 4 the weather improved enough, and the B-17 took off from American soil, headed northeast along the rocky coast to Goose Bay, Labrador.

The next morning in Goose Bay, the weather was good. In Washington, General C. R. Smith was exuberant. He was sure the women would succeed, and he thought a welcoming celebration on the other side of the pond was in order. So he telegraphed Brigadier General Paul E. Burrows in England. As it happened,

Burrows received the telegraph as he was dining with Hap Arnold. Burrows knew of Arnold's work on behalf of women pilots, so when he opened the telegram and read it, he smiled, thinking this was opportune. He handed the piece of paper to Arnold, who put down his fork, and read the telegram. His face reddened. He had approved no such plan, and he knew there would be a tremendous backlash if an American woman were shot down. British ATA pilot Amy Johnson had died in a crash in 1941, and although the cause was never determined, many believed it was from anti-aircraft fire. If male B-17 ferrying pilots needed morale building, they would get it somewhere else. Arnold said, "Ground them. At once."

At about that time, in Goose Bay, the crew boarded the plane, and Nancy in the left-hand seat started the engines and prepared to taxi. Outside on the runway she saw a jeep. A serious-looking young man jumped out and ran toward the plane. It was Goose Bay's commanding officer. Above the engines they could hear him. He was telling them to get out. Nancy and Betty shut down the engines and left the cockpit and climbed down through the forward hatch and dropped feet first to the pavement. The commanding officer handed a radiogram to Nancy. It was from Arnold. It read "CEASE AND DESIST, NO WAFS WILL FLY OUTSIDE THE CONTIGUOUS U.S." They were going nowhere.

A photographer was there expecting to make a record of the takeoff. And because the photographer was there and they were there, Nancy Love and Betty Gillies stood for a photograph, before the plane on the runway with the cold Labrador wind whipping their hair across their faces. It is a striking photograph, especially because Love and Gillies are trying to smile. The two faces in that photograph are the faces of the Trojan women. They are the faces of women who have learned to forbear.*

Jackie Cochran had gone on record saying that individual achievements among WASP were to be discouraged, but it was clear that she had a few of her own. There were, therefore, suspicions that

*Arnold had reservations about exposing an American woman to enemy fire. Nonetheless, many American women were serving in war zones. By 1944, some six thousand nurses would be on duty at AAF station hospitals; and five hundred would be flight nurses, aiding in the air evacuation of the wounded.

Cochran had sabotaged their attempt because she wanted to reserve all "firsts" for herself. Whatever the truth of Cochran's feelings, she knew nothing of the B-17 flight. In fact, at that moment she already had other matters on her mind. In her Pentagon office she had received a memo from Brigadier General Harper of the Training Command: "It is desired that you conduct experiments over a four (4) month period utilizing fifty (50) women pilots for the purpose of target towing, glider towing, tracking, radar calibration flights, co-pilots on aircraft requiring co-pilots." She got to work.

On August 17, two bomb wings from the Eighth Air Force—nearly four hundred bombers—had set out toward the city of Schweinfurt. With no pursuit escort east of the Belgian border, the bombers took the full brunt of German defenses. Sixty B-17s were shot down, and one hundred more were damaged. The Eighth Air Force suspended bombing runs for five weeks. In October, the Eighth Air Force B-17s resumed bombing attacks on Germany. This time, fortune was on their side. On October 4, three hundred and twenty-six B-17s hit Frankfurt; all but twelve returned. But the luck would not last. Over the course of the next week, eighty-eight Fortresses were shot down, and on October 14, sixty were lost in the second raid on Schweinfurt. It became evident that the Eighth still lacked sufficient numbers of long-range pursuits to protect its bombers, and deep-penetration missions were suspended until fighter-escorts with auxiliary fuel tanks could protect bombers all the way to their targets.

Still, the October raids had again shown the B-17 a worthy and capable aircraft, especially remarkable in its ability to endure damage. What was amazing about the raids was not how many B-17s were lost, but how many survived. They flew through flak and antiaircraft fire that took out engines and wingtips and tail sections—and still made it back to their bases in England.

In the fall of 1943, more B-17s would have to be built and more would have to be ferried. And so on October 15, the day after the second raid on Schweinfurt, seventeen graduates from Avenger

Field reported to Lockbourne Army Air Base in Columbus, Ohio, where they would learn to fly the B-17. When the women saw one up close, they were amazed. The Flying Fortress was designed for a crew of nine, and its fuselage was over seventy-five feet long. Its nose featured a huge Plexiglas bubble for the navigator. There was a top gun turret mid-fuselage; a person could actually stand up inside. And mounted beneath its forty-foot wings were four 1,325-horsepower engines. Even empty, the B-17 weighed more than twenty tons.

Flight training in a B-17 meant 130 hours in the air—synchronizing the four engines, and flying and landing when one or two engines stalled, taking off on instruments, and flying to 30,000 feet. It was reasoned that crews would be better able to respond to problems if they understood the plane, so there were ground classes, divided into groups of four. Frances Green found herself with two other WASP and a male lieutenant standing around a table on which was mounted an enormous Wright R-1820-27 nine-cylinder radial air-cooled engine. The instructor in the front of the room told them to take it apart and lay the parts out on the table. That seemed easy enough, and in half an hour it was done. The table was covered with valves and piston rings. Then the instructor told them to reassemble it so that it *ran*. If it didn't, they'd have to do it all over again. Tired and confused, the three WASP looked at the lieutenant, but he just shrugged. Greenie picked up a piston ring, and they began. Hours later, it was assembled. The instructor turned the starter key, and they held their breath. It turned over, caught, and there was the satisfying roar of all pistons firing.

Twenty-four-year-old flight instructor Lieutenant Logue Mitchell wasn't sure why he had been assigned six of the seventeen women students. The major had told him it was because he was happily married, which was true enough. But he had a feeling too that it was because he had a knack for "problem" students.

Mitchell and other B-17 instructors instructed their trainees in "hooded takeoffs" as practice for bad weather flying. The windscreen of a B-17 was divided into two panes, like a 1938 Cadillac, and for a hooded takeoff the instructor put a black curtain over the pane in front of the student. She was expected to get the plane down the runway and into the air without being able to see anything outside.

Greenie was first. She was sitting in the pilot's seat. Mitchell, in the copilot's seat to her right, taxied the plane into takeoff position, locked the rear wheel so the plane wouldn't fishtail, and looked at Greenie. Over the roar of the four engines he shouted, "*She's yours. Relax.*" Greenie swallowed hard. She put her hand on the throttles palm up, and with the heel of her hand gently pushed them forward. The engines grew louder and beneath them they felt themselves bumping. The needle on the airspeed indicator moved past 30, 40, 50. She had to watch the compass to make sure they stayed on the runway. Gently she pushed the throttles a little farther, reaching 80, 90 then, finally, 110, takeoff speed. With both hands she pulled back gently on the control yoke and suddenly the bumping stopped and everything was smooth. She felt a gentle push back into her seat. The airspeed needle was at 135 and the altimeter needle was moving—when it passed 1,000 feet, Mitchell reached in front of Greenie and pulled the curtain off the windscreen. Before and below them she could see the whole city of Columbus laid out like a map.

For Greenie and Mary Parker those weeks were wonderful, and so was that plane. They'd talk about how it felt to fly something that powerful when all four engines were humming. They'd be amazed at how smooth it was, how beautifully it handled, how you could make it do everything you could make a little trainer do—even rolls.

For Lieutenant Mitchell, too, the experience was an education. A few weeks into the course, Mitchell realized he had never had trainees quite like this. He couldn't decide whether they had less understanding of military protocol, or simply less tolerance for it. One day, when Mary Parker was having trouble with instrument flying, Mitchell asked her to surrender the seat to another student. She looked at him and in the even, measured tone of a polite demand, said, "Let me do it again." And he did. These girls were different, he was beginning to realize. Most of his male students did well when they competed against each other, or when he pushed them. But the girls didn't compete. They helped each other. And he soon learned that he didn't need to push them. It was pretty clear that something was pushing them already.

Logue Mitchell was tall, handsome, and—when he was not being an instructor—shy. Sometimes the girls would have fun with this. When they landed at other bases, left the plane and crossed the field to ops, the girls would become a kind of chorus-line escort for their instructor, arms locked and a slightly embarrassed Mitchell in the middle. Ground crews would think they were seeing things. And if, mid-stride, Mitchell dared to voice a quiet reservation about this behavior, one of the girls, through her chorus-line smile, would say, "Kid, you never had it so good."

Mitchell's six students flew through November and into December. Two weeks before Christmas a series of storm fronts came through, and training fell behind schedule. On New Year's Day, 1944, the WASP still needed thirty hours of flying time. Mitchell decided they would do it all at once in a cross-country flight to Texas. And because the winds were calm, they would take off as soon as possible, which meant at night.

Out on the field a light snow was falling, and the base was mostly quiet. As had become standard procedure before a training flight, Mitchell waited in the ready room, sipping coffee and watching his six students through the window. They were doing the walkaround, checking the tires for air pressure, the engines for oil leaks, the brake clearances for wear. Then they opened up the forward hatch, and one by one they reached in, pulled themselves inside, and checked the systems—intercoolers, generators, fuel pumps, hydraulic pressure and fuel transfer valves. When Mitchell saw two of them in the lighted cockpit, he downed the rest of his coffee, zipped his jacket, and ran out across the airfield to the plane.

This time they were going to 30,000 feet. They would wear their fleece-lined leather jackets, their leather helmets, and their oxygen masks. It could get to twenty below zero, so cold that moisture from breath can ice up in the oxygen mask and block airflow. There were low clouds, so Mary, in the pilot's seat, would fly on instruments. Mitchell crouched in the space behind the seats. As they rumbled down the darkened runway and slowly lifted into the air. Suddenly there was a loud, sharp crack outside. It sounded as though they had been hit by artillery fire. Mary and Greenie both turned to look at Mitchell. "Don't worry," he shouted over the engines. "Ice that formed on the props is hitting the wings and engine cowlings. It just *sounds* bad."

At 20,000 feet they broke through the clouds. Through her oxygen mask Mary let out a whoop. Finally she could see out the windscreen. The view from the top turret and especially the view from the navigator bubble was like nothing they'd ever seen. A white moon, riding high, was near full. The cold air was clear, and a long way below them was the cloud deck from which they had emerged. It looked like a great, roiling, silver sea stretching to the horizon. Even the plane itself seemed different. The ice had taken paint off in large patches, and much of the plane's surface was now bright aluminum reflecting the moonlight. It was as though someone had dipped it in silver.

Lockbourne was proud of its graduates, and although training usually ended with a flight check and a certificate, this class was different. For once, they would celebrate their accomplishments in style. A formal dance was to be held at the officers' club in the Fort Hayes Hotel in Columbus. From the street, the hotel was ablaze with light. Inside, in the ballroom, were thirty tables with white tablecloths and on every table was a bottle of champagne in a bucket of ice. There was a twelve-piece orchestra. For a long moment the girls could only stare, and think that the nights on the bunks at Avenger Field seemed like a million years ago.

They danced and told stories about stalls and restarted engines at 3,000 feet and how nervous so-and-so looked on that first day. They sang and they laughed at bad jokes because they didn't want to think about the war. The orchestra played "The GI Jive" and "One O'clock Jump" and 'Don't Sit Under the Apple Tree," and anybody who wanted to dance with one of the WASP had to get in line. But on that night six of those women were saving a dance for one man. And later, with the lights low, each of them took a turn around the floor with flight instructor Logue Mitchell.

By late fall of 1943 the women at Camp Davis had proved themselves so capable that Cochran began to work to get them other assignments. They were sent to transition training in the North

American B-25 Mitchell, and to a highly classified program at Liberty Field in Georgia to fly drone aircraft by radio-control. They began administrative flights, cargo flights and engineering flight testing. At Camp Davis some began simulated strafing flights in which they came screaming out of the sky from 10,000 feet in a steep dive straight at an artillery emplacement. No tow target was used, and no ammunition was used either. Instead, cameras were fixed to the sights of the guns on the ground, and hits were judged later by viewing the developed film. And a few women were transferred to Dodge City Army Air Base in Kansas, where they would transition to a twin-engine medium-weight bomber called the B-26 Martin Marauder.

The B-26 was powered by two 2,200-horsepower Pratt and Whitney engines. It was called "hot," among other reasons, because it had to land at a very high speed—130 to 135 miles per hour—an approach so fast that in any other plane it would be dangerous. The B-26 was difficult also because it required instrument flying—constant, second-to-second monitoring of gyroscope, altimeter, airspeed indicator. For all these reasons a lot of men were calling it "The Widowmaker." In 1942 Senator Harry Truman, observing that more B-26 pilots were being killed in training flights than in combat, had initiated a formal investigation which resulted in some design changes. But even a year later there was, regarding the B-26, what was politely called a "morale problem."

Jackie saw all this as an opportunity. If the boys were afraid of the B-26, she would let the girls fly it—to show them how it was done, and to help them with their little morale problem. The B-26 flight school was in Dodge City, Kansas. Jackie selected the first women candidates herself. And just so they wouldn't overhear the men calling it a widowmaker or a flying coffin or a frigging deathtrap, in a sort of we-have-nothing-to-fear-but-fear-itself policy, Jackie ordered that the women's classes be separated from the men's. In the ground school the women outscored the men on everything—the mechanical system, the electrical system, the hydraulic system, and the emergency system. When the male students at Dodge City found out about this they were flabbergasted. The administration back at Avenger Field received a letter from a Major John Todd which read:

> It is noted with great interest that on the B-26 Procedures Course . . . the average grade of the WASP was 77.8 percent while the average grade of the male students was 74.6 percent. All WASP may well be proud of the high margin of superiority demonstrated in this, the most important ground school course.

And at Avenger Field they also noted it "with great interest," and they were proud of that "margin of superiority." But this was not the kind of thing reported in the military newspaper, *Stars and Stripes*. There was a war on, and news like this sunk like a rock in a pond.

One day at one training base every student was called onto the flight line to watch an airshow involving two B-26s. They stood there and watched these twin B-26s bank and climb and fly chandelles and lazy eights, maneuvers difficult even in a trainer. Sometimes the two planes were only a few feet apart. It was a kind of precision flying the students had never seen. Then the planes were landed hot. The two B-26s, wingtip to wingtip, taxied right up to the students. The engines were shut down, and the props slowed and stopped.

The students were mightily impressed. They applauded. It was like a performance, and now it was time for the curtain call. The figures of the pilots and copilots visible in the cockpits disappeared, and underneath each plane in the nose wheel wells, the forward crew hatches fell open. In the shadow behind the nose wheel of each plane a figure jumped down onto the tarmac. Then a second. And then the four of them came out from behind the wheel assemblies into the sunlight and . . . they were girls. There were four girls standing on the tarmac in front of two B-26s. They were smiling. The students couldn't believe they were the same figures they had seen in the cockpit a few seconds earlier. For a minute they thought it was a trick—like the lady in the magic show who is sawed in half and then wriggles her toes, and the toes are somebody else's. But nobody else was getting out of the plane.

The idea was "*Let's embarrass them into flying it*"—and it seemed to work. The WASP flew B-26s in demonstration flights at airbases across the country, and within the year over one hundred WASP would be trained as B-26 pilots and copilots. Of course, not all the men thought this was a good idea. Many thought that women flying demonstration flights proved nothing. The planes were still dangerous, the women were being used. In fact, when a well-meaning lieutenant thought to make such an observation to one of the WASP, she responded, "Maybe they *are* using us." She looked at him with a patient smile. "But they're also letting us fly those planes."

When the WASP B-26 graduates were sent to various bases for other duties, word of their achievement did not always precede them. A few of the graduates went to Gowen Field in Boise, Idaho, where they were treated kindly enough, but there was still a sort of skepticism. Like a lot of bases, Gowen Field trained gunnery crews and pursuit pilots in target practice. Because the B-26 was fast and maneuverable, some of those at Gowen Field were stripped of armament and used for high-speed target-towing.

Pat Patterson, in the left-hand, pilot's, seat and Marge Gilbert, in the right-hand, copilot's, seat, were flying a B-26 tow-target, when at 10,000 feet the accessory drive shaft in one of the engines fractured. They were suddenly flying with only one engine. To keep control of the plane Pat pulled back on the throttle (the thrust from a single engine is off-center, and causes a plane to yaw). At the same time, with her foot, she hit the rudder pedal hard and held it there. She put in the requisite call to the tower, and the tower told them they were number one to land. Using the rudder to compensate for that one engine giving off-center thrust, she descended to 2,000 feet. Marge to her right prepared to lower the landing gear which ran on a hydraulic system—and discovered the hydraulics weren't working. So the flight engineer sitting behind them began to crank the wheels down by hand, and because this took a while and they didn't want to have to turn the beast again on one engine, they had to try for the next runway, but the next runway was "closed"—planes were parked on it. There was no other choice. If they were going to get down at all they would have to land on a runway full of parked planes. There was a narrow corridor of open pavement between those planes. But with the thrust still coming off-center, they knew that the moment all the wheels were on the ground it would not be like flying a plane anymore; it would be like driving a fully loaded freight truck on a narrow stretch of road. Except that this truck would have blown all its tires on one side and would be moving at an approximate speed of 130 miles per hour.

By this time word had spread through Gowen Field that the plane in trouble was piloted by two WASP. And everybody—the mechanics, the other male pilots, the staff, the other WASP—had stopped whatever they had been doing to watch. The plane was badly crippled, and its pilot and copilot were about to attempt something which had never been done and which some of those watching would have said was impossible.

The B-26 in the distance got lower and lower and seemed to be moving faster—even on one engine it had to hit the runway at 130 miles per hour or it would stall, and holding it at that speed was hard enough when two engines were working. It was going in fast but true, nose up, flaps down. The only sound was that single engine. Then the rear wheels hit and there were two little sprays of dirt and then the nose came down and the nose wheel touched and all wheels were on the ground and the beast moved on through. It was rumbling along the runway through the narrow open corridor between the parked planes . . . and the girls were . . . holding it straight . . . and it stayed straight . . . and it slowed and . . . slowed some more and . . . stopped. For a moment there was just quiet. Then somebody let out a whoop. And then another. Soon there was cheering from the boys out on the flight line and the boys outside operations and the boys in the control tower and the boys out on the ramp.

The class of 43-7 graduated on November 13, and within days its fifty-nine members were settling into accommodations at air bases around the country. Only a few were sent to ferrying bases. Cochran's promise that the women could fly anything the Army built had been realized. Some WASP were giving check flights to veterans being introduced to newer versions of B-24s. Some were flying war-weary B-17s, just back from Europe, to plants where their aluminum would be recycled into B-29s. Some were flying C-30s, towing the enormous gliders that the U.S. infantry would ride above the beaches of Normandy. And that November, nineteen members of that class had been assigned to Mather Air Force Base, where they were to learn to fly the B-25 Mitchell.

The women had heard that there would be black pilots there. The AAF had been conducting training for all black pilots at Tuskegee Field in Alabama, and by June, graduates who formed the 99th Pursuit Squadron had already seen action in North Africa. In late 1943 the AAF began plans to form the 477th Bombardment Group to fly twin-engine bombers. Because Tuskegee had no facilities for bombers, it became necessary to send students to other fields for transitioning—Hondo Field in Texas, Midland, Texas, and Mather Air Force Base in Sacramento.

All trainees, regardless of skin color or gender, dined at the same mess. There were a great many black men in mess at Mather Field, and there were only nineteen women, yet every night the women were served first. The women began to think this was strange. They and the men were taking the same classes, they and the men were flying the same planes. So one evening one WASP said to the officer in charge, "We've talked this over, and really there's not a single reason the WASP should be served first." He agreed. The next night servers started from the other side of the room. There were smiles and nods. So began a brief friendship of a remarkable group of women and a remarkable group of men, both of whom the AAF regarded as experiments.*

In 1922 an Army Air Corps lieutenant had been saved by a parachute, and his fellow officers had presented him with a little piece of paper they called a "caterpillar certificate." The name was a

*The 477th would be plagued with problems stemming from racist officers and mismanagement; consequently, the black graduates of Mather Field and other twin-engine bases would never fly a mission. Some of their members would, however, enter struggle with a nearer enemy. On April 1 and 2, 1945, fifty-seven black officers stationed at Freeman Field in Indiana entered a base officers' club as an act of civil disobedience. All were arrested and confined to their quarters. The 477th commanding officer, acting on the advice of the judge advocate general of the Army Air Forces, released all but three who had been charged with using force. In July, two of the officers were acquitted of the charge of using force, and a third was sentenced to a forfeiture of fifty dollars per paycheck for three months.

Meanwhile, the all-black 332nd Fighter Group, based in Italy and flying missions into southern France, Austria, Yugoslavia, Greece, and Germany, would distinguish itself in combat from January 1944 until the end of the war. Its pilots would be credited with destroying 158.5 enemy aircraft, and on escort missions they would not lose or abandon a single bomber. (The number after the decimal was the AAF's indication of a "shared kill.")

No WASP or WAFS were black, although Cochran would later say that there had been at least one qualified applicant, whom she, Cochran, dissuaded from entering training for her own safety. Women ferry pilots often had to travel alone to airbases that could provide them no overnight accommodations, and in the U.S in the 1940s, few hotels or rooming houses would accommodate blacks.

reference to the silk of which his parachute was made. Bailouts happened often enough that this developed into a little tradition. If a pilot bailed out of a plane and kept cool enough to have the handle on his parachute release ring when he was standing on the ground, he could call himself a member of the Caterpillar Club. In fact, this lighthearted tradition belied a rather serious situation. After the war a study would be made of twenty-five hundred emergency bailouts between 1943 and 1945. The sobering conclusion was that 58 percent resulted either in serious injury or death. During the war, however, bailouts were a last resort, and at Avenger field almost no attention was given to them. Trainees were shown an Army instruction manual with cartoon figures.

Trainee Lorraine Zillner was in WASP Class 44-W-2. Lorraine was small and she looked younger than she was. When she wore her hair in braids she could have passed for fourteen or fifteen. By November 1943 Lorraine was already a pretty good pilot. She was the first of her class to solo in the PT-19, and like everybody else she was thrown into the Wishing Well. On the whole, Lorraine was enjoying life at Avenger Field.

One morning Lorraine was practicing aerobatic maneuvers in a BT-13. The BT-13 had a 440–horsepower engine and a lousy reputation. Those supplied to Avenger Field were old. They rattled, the cowlings came loose, the radio reception was bad, and in general they were what was politely termed "unreliable." The girls were not as kind. They called the planes "those buckets of bolts" or, if they were in a darker humor, "Bunsen burners."

Lorraine was beginning a turn when the plane went out of control and flipped into an inverted spin, what the flight manual defined as "the flight condition of an aircraft which is flying upside-down with regards to the ground and which is descending vertically, circling the vertical axis of descent." Lorraine was suddenly hanging upside down from her seat belt; she could see the ground over her head where the *sky* should have been—and was spiraling down toward it too fast. With her foot she hit the rudder pedal against the direction of spin. The plane didn't respond, so she eased off on the rudder pedal and there was no difference. She tried again, hoping maybe something would catch, but it was like hitting a brake pedal in a car when there's no brake pressure. The ground over her head just kept getting closer and she knew she had to get out. But she was hanging upside-down and the blood was rushing to her head and the whole earth seemed to be turning above her. She knew that if she jumped the slipstream would pull her back toward the tail and because the plane was upside-down it would be hard to avoid hitting it. There was a cotton field beneath her and it was getting closer, so with one hand she grabbed the edge of the cowling and with her other hand she released her seat restraint and then let go of the edge of the cowling, kicking hard against her seat at the same time. The slipstream pulled her back along under the fuselage, and she was like a rag doll in a hurricane. Her leg smacked hard against the tail and she actually bounced off the plane. At least now she was falling away from it, but she was also tumbling and she'd been trained to count slowly to ten before pulling the ripcord. She thought *to hell with the training, that cottonfield is too close* and so she counted fast, "One ten" and pulled hard. The chute fluttered open and held and the harness straps jerked her upright and . . . she hit the ground.

For what seemed like forever she lay there among the cotton rows. It was quiet. Her leg hurt. Blood throbbed in her ears. Then through the ground she felt horses' hooves and then she heard men's voices. One was saying, "There's the pilot." They were cowboys, two of them, and they came galloping up and got off their horses and knelt down near her and one of them helped her get her leather helmet off. Her hair came tumbling out and the other cowboy said, "My gosh—it's a *little girl.*" And for some reason this was just too much—the crashed plane and the training and her leg hurting like hell and the shock of the jump and then these *cowboys*. She started to cry. One of the cowboys said, "Don't cry—we'll get help." The other one was holding a branch of cotton in front of her face and telling her she could use it to dry her eyes.

Back at the infirmary her leg was bandaged up. It was only bruised. A few days later she appeared before the board for questioning in the style of a formal military inquiry. Lorraine was asked why she had abandoned her airplane, to which she replied that she had stayed until the last possible second. After a few minutes it was established that she had done all she could do to maintain the integrity of her airplane, and that she had in fact delayed too long before abandoning it. Otherwise, she acted in a manner entirely consistent with her training. The other girls had been particularly impressed to hear that she had been holding the parachute release ring when the rescue team carried her into the infirmary. Then one of the officers asked what they were all thinking, "*How did you manage to get out of a plane that was in an inverted spin?*" Lorraine looked at him and said, "Sir, have you ever prayed?" Nobody could think of anything else to say, and the presiding officer ended the proceedings.

15

LADIES OUTRAGEOUS

In December, the Ferrying Division opened a pursuit school in Palm Springs. Fifty-six women were listed for pursuit transition. The newer version of the P-51 Mustang had improvements—new cameras, and gunsights and IFF (Identity Friend-or-Foe) sensors and transmitters. And the pilots, as part of their flight gear, were given a .45-caliber pistol—not to protect themselves, but to protect the design changes from spies. In the event that the Mustang was forced to land offbase, they were taught to fire at a place on the fuselage holding explosives.

One of the graduates was Florene Miller. She had black hair and great dark eyes and skin like china. Even wearing the bulky fleece-lined leather flight suit, she could stop traffic on an airfield. Florene took some time with her appearance, and when she was transitioned to pursuits this became a challenge. Pursuits had small cockpits, and once she had the parachute behind her and the forms stuck in the holster on her left, the emergency medical kit and the .45 packed away, and especially if she was wearing winter flying gear, there was only enough room left over for a little canvas pouch that would hold a change of underwear and a toothbrush. But Florene was not about to neglect her appearance simply because aircraft designers had lacked the foresight to accommodate the needs of a modern American woman. She used to say that after a long mission she wanted to "turn into a girl again." Florene had with her makeup and a mirror and a pair of dress shoes and perfume and at least ten shades of lipstick. She would find places all over the plane to store these things. She carried a little screwdriver in her purse, and she'd open a panel in the wing and tuck a shoe into an ammunition box between two internal braces, then find the corresponding panel on the other wing and tuck the other shoe in that ammunition box. Florene Miller knew more about the nooks and crannies of the P-51 than its builders.

Although the styles of Florene Miller and other American women were not abandoned for the necessities of war, they were much changed. The Manhattan fashion establishment, lacking the examples set by Parisian designers, turned for new ideas to the military. And by 1944 it was not uncommon for a well-dressed American woman to walk to work wearing an imitation British commando beret, or to attend a concert in a wrap with a drawstring waist modeled on an Eisenhower jacket, or to wear an imitation WAC hat trimmed with sequins. A popular evening gown of the period was adorned with a pair of great Army Air Force wings of gold lamé, swooping from hip to shoulder.

But for all the unintentional influence military dress had on civilian fashion, in spring of 1944 the AAF had yet to recognize the sartorial needs of many of the women flying its aircraft. Nancy Love's original WAFS had worn uniforms of gray serge, including slacks, a jacket with the arm patch of the Air Transport Command, and the wings of ATC Service Pilots. But WASP were wearing provisional and undistinguished uniforms of tan gabardine slacks, white shirts, and overseas caps. On ferrying missions, problems of recognition were many, and a few had been arrested for impersonating an officer. For months, Cochran and Arnold had been considering an official WASP uniform, and now the AAF offered them a choice: that they adapt uniforms that had been rejected by the Army Nurse Corps, or design new uniforms using excess WAC material of Army green.

Jackie Cochran knew that the uniform would mean everything for the girls' morale, and for the way they'd be treated by others. It was no small matter. Jackie was aware that fashion of recent years had been influenced by the Army, and saw no reason the direction of this influence might be reversed. Cochran hired a Manhattan designer named Bergdorf Goodman. Mr. Goodman got to work, and one day in the Pentagon Cochran arranged a kind of fashion show, with Arnold and Army chief of staff George C. Marshall as its audience and its judges.

First an Army typist entered Arnold's office. She was wearing a version of the WAC uniform, whose color, officially, was kindly termed "khaki." It was the color of newspaper left in the sun for a month. Then a second Army typist entered, wearing the Nurse Corps uniform. It was difficult to distinguish it from the first. And then in that airless Pentagon office it was as though someone had suddenly thrown open a window, because walking confidently into the room was a tall, slender woman. She was smiling a white smile. And she was wearing a beret and a strikingly handsome, perfectly tailored blue suit. It was fitted about the waist with a serge-belted jacket. The beret and suit were a dark shade of blue they called "Santiago." And this woman walked across the room, turned once and smiled that white smile. It was obvious she was a professional model.

Arnold knew that Jackie had stacked the deck, and Arnold knew that Jackie knew that Arnold knew. Hap Arnold put in his vote for the Santiago blue anyway. On February 11, 1944, the forty-nine graduates WASP Class 44-W-1 graduates were issued uniforms.

In mid-February, the Eighth and Fifteenth Air Forces concentrated attacks on the German aircraft industry for six days. It was called Big Week. Over three thousand Eighth Air Force bombers and five hundred from the Fifteenth dropped more ordnance on German aircraft factories than the Eighth had delivered in the previous year. There was a second success, perhaps more significant: The operation involved the first large-scale employment of P-51s with wing tanks. It was clear that escort pursuits could protect Allied bombers ranging over the entire Reich. In Washington, Hap Arnold knew that American airpower had come of age, and the airwar over Europe had entered a decidedly new phase.

Meanwhile, U.S. civilians bought war bonds, collected scrap metal, and planted victory gardens. Many worked in industries which supported the war effort. Even Hollywood was making training films for the Army. In early 1943 Universal Pictures made plans to film the story of the WAFS. Universal had requested Nancy Love's cooperation, which she treated as per ATC recommendations, with diplomacy and delicacy. Still, she asked that the film be factual, and they assured her that it would be. Universal needed an actual plane, a PT-19, and Teresa James was assigned to deliver it to California. When she arrived the producer arranged private tours of the studios, and a few movie stars asked for her autograph. At the famous movie star restaurant the Brown Derby, Teresa had lunch with Mr. Bob Hope.

By April 1944, only a few WASP remembered that a film was being made. They had other concerns. Chief of air staff Barney Giles wanted all the WASP to attend officer training, and on April 9, Cochran sent fifty to the AAF Strategic Command School in Orlando. The first class was made of squadron leaders and women who had been on operational duty the longest. Nancy Love, Betty Gillies and several of the original WAFS attended—it was a kind of reunion, the first time they had been together in more than a year.

For four weeks they studied aircraft recognition, advanced meteorology, and military law. They studied eight hours a day, six days a week, and they were tired at night. Nonetheless, when a film called *Ladies Courageous* appeared in an Orlando theater, a few WASP and WAFS went to see it. The opening credits said that the film was "sanctioned by the United States Army Air Force as the official motion-picture story of the Women's Auxiliary Ferrying Squadron, now known as the WASP, Women's Air Force Service Pilots." Loretta Young played a beautiful, strong-willed woman called "Roberta Harper." Roberta Harper divided her time between fighting Washington bureaucrats and sorting out the personal problems of her young charges. A high-strung young pilot named "Virgie Alford" crashed a plane deliberately, to get attention from the press. A vixenish WASP named "Nadine Shannon" flirted unashamedly with the husband of another pilot. When the pilot discovered this indiscretion she became depressed and angry enough to commit suicide. It was a rather spectacular suicide, and for the Army Air Force a very expensive one: she used a P-40 Warhawk.

The WASP began to wonder how this film about a group of hysterical women could possibly have been "sanctioned by the United States Army Air Force." By the time the women returned to barracks that night they were calling it "Ladies Outrageous."

The film was melodrama whipped up to a sensational froth, yet there was at least some truth in it. At certain levels of the ATC the WASP were gaining a reputation as careful flyers, the ones who were not about to try hassling and the other cowboy stuff—the strafing runs on farmhouses, the fence-jumping, the flying upside-down, the buzzing main streets of small towns. But it was also true that high-altitude high jinks were not exclusive to the male of the species. The WASP, after all, were healthy young women, far from home, and maybe for the first time in their lives the world was full of possibility—and the sudden end of possibility. In fact, their reputation as cautious was only partly deserved.

But they did have fun. On one occasion two Navy combatants practicing aerobatics in the airspace near California's Salton Sea heard a third voice in their headphones. It was a woman's voice—"*Red Dogs One and Two this is . . . uh . . . Red Dog Three . . . and right now I'm waxin' both your tails.*" It was WASP Helen Turner. The tradition was that, after a bit of hassling, the faux combatants would fly alongside each other just long enough to see each other's faces. Maybe they'd give a thumbs-up, in a little bit of World War I fighter ace chivalry. Helen was ready to participate in this custom too, but because she was flying a Mustang she had to lower flaps and actually slow down to let the two Corsairs catch up. And when the Navy aviator saw it was a woman in that P-51 cockpit, he thought, *man the Army sure is changing*.

That evening at the base there was a phone call for Helen. She was a little worried somebody had reported her, and so when she answered the phone she was wary. But it wasn't a commanding officer—it was a Navy aviator. In fact, it was the one whose tail she had waxed that afternoon, and he was asking for a date. "How on earth did you get my name?" she asked. Through the phone she heard him laugh. "It wasn't hard. These planes have tail numbers."

All this might be what a psychologist called sublimation of

youthful impulses. Except they weren't always sublimated. America at war allowed for the relaxation of social strictures. If a boy was leaving on the next troop transport for, say, the Eleventh Air Force in Alaska, or a carrier in the Pacific, he would remind a girl that he might not make it home. There was the exuberance of youth, and there was the romance of young men fated to live or die in an exotic place halfway across the world. Throw airplanes into the mix and you had carpe diem to spare.* The WASP made these *unsublimated* impulses into a sort of institution. It was called the "Mile-High Club." Membership did not discriminate against race, creed, or place of national origin. Neither did it discriminate against gender. There was only a single requirement, and it was utterly straightforward: one had to have made love at or above an altitude of 5,000 feet.

There would be opportunities. WASP were at first prohibited from catching rides with men, as part of Cochran's plan to keep up appearances. But male pilots could hitch rides with WASP. Some of the transports and medium and heavy bombers required two people to fly them, and many ferrying crews were of mixed gender. A plane like the B-17, for instance, offered areas of privacy. The midsection was some distance from the cockpit, and with four engines running there was no possibility of eavesdropping. Or, if one preferred a romantic panoramic view of the world passing 20,000 feet beneath, there was the navigator bubble. . . .

By the beginning of 1944 Hap Arnold considered the Army Air Force's experiment with women pilots a success. As a group, the women had not only outperformed their male counterparts in ground school exams and flight tests, they had also proved themselves more willing to tackle any assignment—no matter how dangerous, difficult, or dull—with efficiency and without complaint. There had been some worthy individual achievements, too. In the summer of 1943, B. J. Erickson had made four two thousand-mile deliveries in six days, a feat for which she was awarded the Air

*The attitude would be particularly noticeable to the British. In 1943 the British welcomed the American pilots of the Eighth Air Force. But after a while they noticed that the Yanks were rather rambunctious, and some of the locals began to call them "overpaid, oversexed, and over here."

Medal, thereby becoming the first civilian to receive the honor since Amelia Earhart. By January 1944 Arnold decided that the WASP had endured their quasi-official, stepchild status long enough. He was prepared to do everything in his considerable power to make them legitimate.

At first it seemed as if it would be easy. The War Powers Act of September 1941 had made provisions for the direct commissioning of temporary officers into the Armed Forces in time of war, granting them the same status and benefits held by career military men of the same rank. Male pilots had been brought into the ATC under these guidelines; it seemed the most expedient way to militarize the women. But on January 13, the deputy chief of air staff responded to Arnold's query saying, "The authority in the act of September 22 [1941] to make temporary appointments as officers in the Army of the United States . . . refers to and contemplates *men* exclusively, and may not be regarded as commissioning women as officers in the Army of the United States." There were no existing channels through which the Air Force could commission women. To militarize the WASP would require an act of Congress.

In January 1944 Representative John Costello introduced H.R. 4219, authorizing AAF commissions for women pilots on duty, granting them all the privileges, insurance, hospitalization, and death benefits accorded male Army Air Force officers. And on March 22, Arnold, accompanied by Cochran, Ethel Sheehy, and Nancy Love, appeared before the eighteen members of the House Military Affairs Committee to ask for their support of the Costello Bill. Secretary of War Henry Stimson had written a letter for insertion into the record, offering his approval on behalf of the president.

Arnold had appeared before the committee numerous times since the war began; usually they gave him what he needed, and they were fully prepared to do the same today. The present situation seemed different, however; members of the committee had been besieged by letters and telegrams from civilian pilots asserting that Army Air Force positions that were rightfully theirs had been usurped by women. Could Arnold explain this? He could. In January the need for pilots had nearly peaked, but at the same time the War Department had increased its estimated need for ground troops. Already some thirty-five thousand Army Air Force officers on the waiting list for flight training had been transferred to the infantry. In fact, with no need to train more pilots, Arnold had deactivated the pilot training activities that had been the Civilian Pilot Training Program before the war, and since the summer of 1941 had been redesignated the War Training Service. Arnold's action put ten thousand instructors out of work. As instructors, the men had held reserve status; now they were eligible for the draft into the infantry. These were the men who were lobbying Congress.

Arnold assured the committee that there was ample room in the Army Air Force for any and all of these men, as ground crew or navigators or gunners. But physical standards for civilian flight instructors were considerably less stringent than for Air Force pilots, and many of the men were simply unqualified. In closed session with the committee, Arnold suggested that the men's readiness to complain suggested they were ill suited to serving in his Air Force. By way of contrast, he said, the attitude of the WASP was exemplary; they routinely performed duties at which even his best pilots balked. Within hours the committee announced its full support for the Costello Bill, which it would send to the Rules Committee to be placed on the House calendar. Hap Arnold was confident that the matter had been resolved.

16

SPRING, 1944

Through the train window Nancy Batson looked out over the Nebraska fields. There were in some places small shoots of bright green—wheat, she supposed. She looked out the window at the greening fields going by and could not quite believe it was already spring.

Batson was acting as a uniformed representative delivering the body of fellow WAFS pilot Evelyn Sharp. She had a ticket for herself and a ticket and a baggage check for the coffin. It was her responsibility to manage all transfers en route, and get off the train at each stop to look into the baggage car and make sure the coffin was still there.

On April 3, Sharp had taken off in a twin-engine P-38 when one engine failed. A lesser pilot would have spun in immediately, but Evelyn managed to keep the nine-ton fighter in level flight, and then managed a 180-degree turn back toward the airfield. But on final approach the landing gear would not come down, and she had no choice but to try to "pancake" the plane—that is, stall it deliberately a few feet above the ground so that it fell the last few feet. She managed this too, but the forward section of the plane hit the ground first, and when the plane jerked suddenly to a stop

Sharp's own inertia threw her upward against the closed canopy, snapping her neck and killing her instantly.

Although Evelyn Sharp had been flying a military plane for the U.S. government and the war effort, and although she had sacrificed her life to that end, she had not been granted a military funeral. In fact, as a WAFS pilot, she did not even qualify for Army death benefits. And so the WAFS themselves took up a collection to transport the body. Betty Gillies, who insisted that the body be accompanied by a uniformed escort, had paid Nancy's train fare.

The people of Ord, Nebraska, were proud of America and proud of Evelyn, and if they were told the government said no military funeral, they may not have understood its reason, but they would have supposed there must be a good one. But at the wake, Nancy realized that no one had told them. A worker in the funeral home saw Nancy's uniform and asked her if he might be allowed to drape Evelyn's casket with the American flag. Nancy knew that this would be an improper assumption of Army protocol and realized this man was innocent of such protocol. She thought of the rationales of the AAF disallowing the women benefits accorded men ferrying pilots, and she felt a righteous white-hot anger. But she did not let the man see this. Instead she looked at his face and managed a smile. Then she took his hand and pressed it tight, once. And she said, "*Of course*."

The next day, at the funeral, Nancy greeted the family and the friends and she told them that Evelyn made her proud to be a WAFS pilot and that they could be proud of Evelyn's accomplishments. But she said it to so many people that she assumed there was a mistake; there were so many in the cemetery that at first she thought there must be another funeral. It was as though the entire town of Ord was there. Then it became clear that there was no mistake. It was all for Evelyn.

The ceremony began and there were so many eulogies that the service took two and one half hours. As Nancy listened she learned that Evelyn had been Ord's favorite daughter. Evelyn had given their children swimming lessons, and had flown their first sack of airmail. The town was so proud of her that the local businesses bought her an airplane . . . and when Evelyn began to fly airplanes for the war effort they were bursting with pride. The mayor was the last to speak, and he announced that the town airport would be renamed Evelyn Sharp Field. Through all this Nancy conducted herself with dignity. When the ceremony ended, the casket was lowered, and the mourners left. And Nancy sat alone among the empty folding chairs and cried.

Later, Nancy was sitting on the train again with Ord already miles behind her. It was early evening and again she looked out the window at the greening wheat fields going by and again she could not quite believe it was spring. And suddenly she realized that she could not believe it was spring because she could not believe that she had been flying with the WAFS for only nineteen months. Soon it was dark enough outside that she could see her reflection in the glass. She looked at her reflection and thought, *I have been flying with the WAFS for nineteen months and I am twenty-four-years old and already I am burying friends I haven't had time to know.*

April would be a bad month. Before it was over, four more WASP would die—in Sweetwater, two on a single day. On April 16, trainees Mary Howson and Elizabeth Erickson, flying an AT-6 and a PT-17 respectively, suffered a midair collision over a runway. As had become standard practice, while the rescue crews did their job, and an investigation board began its inquiry, Mrs. Deaton arranged for transportation of the bodies. She had been in her position now for more than a year, and on good days she was amazed at the girls. They seemed to think of nothing but flying, and sometimes she would joke that they had "propellers in their heads." But she could not accustom herself to their morbid sense of humor. They had covered a wall in the Link Trainer room with small funeral wreaths with the names of Link students who had "crashed." And when a girl would climb out of a plane she'd groundlooped or braked so fast it had nosed over, her friends, upon seeing her walking back to the field, would shout and laugh and greet her with a chorus of "Blood in the Cockpit."

It had gone on for more than a year now, and still Mrs. Deaton was not used to it. And she could find nothing that would help her to *get* used to it—death was a subject little discussed in the officers' guide. Cochran had said she was supposed to "watch over the girls," and she did that and she would continue to do that, but the assignment had made it sound as if she were a housemother at a college sorority where she might have to deal with, what, homesickness, a bout of flu, maybe a little sneaking over the wall at night? What sorority housemother had charges who were taking fragile machines to 10,000 feet and deliberately putting them into spins and stalls and dives? What sorority housemother had charges who were risking their lives in a hundred ways on a daily basis? And what sorority housemother was expected as part of her duties to care for the body of a young girl—a body which had been crushed by an airplane?

When she had arranged for transportation of the bodies of Mary Howson and Elizabeth Erickson, Mrs. Deaton sat alone in her office, leaned forward and crossed her arms on her desk. Then she buried her face in her sleeve.

In England, the Eighth and Ninth Air Forces gathered pursuits and medium bombers, and joined their forces with the RAF's Tactical Air Force in preparation for the retaking of the continent that would be called Operation Overlord. If they could isolate the Normandy region, they would prevent Germany from bringing up reinforcements once Allied troops hit the beaches. In May, the American bombers turned to transportation targets in northwestern France, especially rail lines. They also destroyed all but one of the twenty-two bridges across the Seine River on main roads between Paris and the Channel coast. German supply and reinforcement now moved only at night, and over circuitous routes.

The Pacific Theater was another matter entirely. Any military advances there would have to wait until the successful deployment of a long-range bomber. With astonishing foresight, Arnold had envisioned such an aircraft before the war began. The AAF contracted Boeing Aircraft in Seattle for the project, and by the fall of 1942 two prototypes were ready for testing. They were called XB-29 Superfortresses. The *X* meant "experimental." The plane was astonishing. Its aerodynamics were refined, it had four turbo-supercharged Wright engines, an electronic fire control system, and forward and aft cabin pressurization. It also had a range of over 5,500 miles, a service ceiling of 30,000 feet and a maximum airspeed of 360 miles per hour. But most impressive was its size.

It was a third again as large as the B-17: The Superfortress wingspan was 140 feet, and it was three stories tall. Next to the Manhattan Project, the B-29 was the most expensive single project in the entire American war effort, and its outcome was every bit as uncertain. Even K. B. Wolfe, the man who was the AAF director for the program, called it a "three-billion-dollar gamble."

In the first months of 1943 Boeing was testing prototypes. The pilot Boeing chose to perform most of the flight tests was named Eddie Allen. Allen was a test pilot's test pilot, the best in the country, maybe the world. He had flown for all the major manufacturers—North American, Lockheed, Douglas, Grumman—and in 1943 Boeing counted itself lucky to have him.

On February 18, Allen, a test pilot named Bob Dansfield, and a crew of nine flight test engineers took off in a prototype to evaluate the powerplant cooling and propeller operation. At 5,000 feet the number one engine caught fire, and Allen turned back toward the plant, radioing in his condition. At that moment Boeing executives were in a weekly meeting, and when somebody rushed into the room and said that Eddie Allen was bringing back a Superfortress with an engine on fire, they adjourned. In moments the vice presidents and assistant vice presidents were all out on the pavement in front of the headquarters. As the plane passed over Lake Washington Bridge a fire began in the number two engine, and soon it was streaming thick, black smoke. In the preternaturally calm voice of the test pilot, Allen radioed back, "Have fire equipment ready. Am coming in with wing on fire." The plane was just north of the field, over Seattle's business district, when the pilots lost the struggle. The fire spread to the fuel tanks, and the plane erupted into flames and fell earthward.

Allen was posthumously awarded the Daniel Guggenheim Medal, and Boeing named its new research facility in his honor. A congressional investigation determined that the engine problems were caused by poor quality control at the manufacturer. For a while the B-29 became known as the plane that killed Eddie Allen, but development proceeded apace.

The Twentieth Air Force was activated on April 4. Although it would operate from bases around the Pacific, Arnold himself would act as its commander. Part of its mission would be the strategic bombardment of Japan. Hundreds of B-29s were flown from a base in Wichita, Kansas, across the Atlantic, across northern Africa and then on to bases in India. On April 13, there were seven wrecks and accidents. The rest were grounded, and Wolfe told Arnold that it was "imperative that improved engine cooling be obtained immediately."

The original design had been an attempt to build an engine with less than one pound of weight per horsepower, and that made for a very hot engine. The engineers knew that a redesign was the only way to stop the overheating, but there was no time, so they worked around it, building in ducted baffles, shortening the cowl flaps, using engine oil for cooling. Still the engines tended to overheat. And again there was, among the pilots of the AAF, a "morale problem."

One evening in May 1944, Dorothea Johnson was leafing through a magazine in the lounge of the nurses' barracks at Eglin Air Force Base in Florida. She was twenty-five years old, and she had recently been transferred from tow-target duty at Camp Davis. Her friends called her DeDe.

A young lieutenant colonel walked in, and DeDe, looking up from her magazine, assumed he was some nurse's boyfriend, here to pay a visit. So she returned to the magazine. But he stood in front of her, and without ceremony, asked, "Do you have any four-engine time?" Again she looked up, this time a little startled. He was a lean man, with dark, intense eyes and thick black brows, and he obviously knew what he was here for. Then he seemed to realize he had caught her off guard, and he backed up a little. "I'm looking for two WASP to check out in the B-29."

DeDe thought he must have misspoken. For a moment she had to repeat his words in her mind to be sure. The atmosphere in that room had suddenly become military, and she found herself standing. She didn't know whether she should stand at attention, but she did. And then she found her voice. "I have a little time in a twin-engine trainer, sir. . . ." She thought maybe he had been given wrong information. There were a few WASP at Eglin, and none had much more experience than that. She began to speak rapidly. "That's all any of us have, sir. Except Dora. She's checked out in the A-20. She's up there now." As they both knew, it had only two engines. DeDe stood there and thought that would be the end of it. But he said: "I'll take her. And you. Do you want to do this?" DeDe caught her breath. She couldn't quite find the words, so she just looked at him and nodded. He said he was Lieutenant Colonel Tibbetts. And she should be ready in the morning. He'd find Dora himself. "And, Miss Johnson?"

"Sir?"

"WASP are Civil Service."

"*Sir?*"

"You don't need to stand at attention." He broke into a smile. "Relax."

And then he was gone.

The experiment began at an aircraft modification center in Birmingham, Alabama. Months of training were standard for transition to any kind of four-engine, but Tibbetts expected Dora and DeDe to do it in three days. They learned emergency procedures, they learned to fly with one engine inoperative and with two engines inoperative, they learned to cruise at various speeds, to glide, to climb, and to recover from stalls. And just so the Training Command could be sure Tibbetts wasn't overestimating the women's abilities, another instructor administered the check ride. Then it happened—an engine caught fire. Dora and DeDe cut fuel to that engine and, without making a fuss, landed the plane. By then most of the test was accomplished, and another check was deemed unnecessary.

The plane would be used only in demonstration flights. They named it *Ladybird*. With a crew of five, and Tibbetts crouching between their seats, the two WASP flew *Ladybird* from Birmingham, west across Mississippi, Louisiana, and Texas, to Alamogordo, New Mexico, where there was a base for very-heavy bombers. Again the men were called out onto the flight line to watch as a plane flew with three engines and two engines, and again the plane landed and slowed to a stop and women dropped out of the forward hatch onto the tarmac.

Chief of air staff Major General Barney Giles was not so much against women flying a dangerous airplane as he was against them *dying* in one, especially in front of crews in training. He knew it was dangerous, and was certain that sooner or later Tibbetts's morale-raising exercise would backfire. Two days later Dora and DeDe were sent back to Eglin. But already Tibbetts noted a change in the men's attitude. However brief an exercise, it had worked.

Meanwhile, the Costello Bill was slated to come up for a vote in June, and considering the gathering of forces in Europe, it ought to have been one of those minor congressional proceedings that attracted little public attention. All it asked, after all, was that some one thousand Air Force personnel have their status changed from civilian to military. But again it did not prove to be that simple.

In March, the House Committee on Civil Service, chaired by Representative Robert Ramspeck, launched an inquiry into government spending. The WASP status as civilian employees of the Army Air Force placed them squarely within the Ramspeck Committee's jurisdiction. In part because 1944 was an election year, the committee was particularly responsive to constituents. The Ramspeck Committee, like the House Military Affairs Committee, was receiving letters and telegrams from civilian male pilots, recently in the employ of the War Training Service, now objecting to the WASP. An investigation of the budget for the WASP program, precisely *because* that budget was relatively minuscule, would demonstrate the committee's vigilance.

In the spring of 1944, a flurry of editorials about the WASP appeared in many of the nation's leading newspapers. *The New York Times*, *The Boston Globe*, and *The Herald Tribune* supported militarizing the WASP; *The Washington Post*, *Time* magazine, and the *Daily News* said it was time for the women to step down and let the men take over. Among those demanding a WASP deactivation was Drew Pearson, author of a widely read and influential column in the *Washington Times-Herald*. Over the next few months Pearson devoted several whole columns to the WASP. He suggested that the use of Civil Service funds to train military pilots was illegal. He made personal attacks on the director of women pilots, insinuating that Arnold had been seduced by Cochran's feminine wiles: "She has even persuaded the Air Forces' smiling commander to make several secret trips to Capitol Hill to lobby for continuation of her pets, the WASPs." The War Department ordered that neither the Army Air Force nor the WASP themselves should respond to Pearson's remarks, or to the charges levied against them by the male pilots, lest anyone inadvertently provide fuel for this anti-WASP fire.

Meanwhile, spurred on by Pearson's sympathy, the male civilian pilots aligned themselves with veteran's groups and stepped up their letter-writing campaigns. Congress got yet another earful of their sentiments. And on June 5, the Ramspeck Committee presented the results of its inquiry. The WASP program so far had cost $50 million, a figure that should have required specific legislative action. Trainees who failed to graduate were a loss to the taxpayer because, unlike male cadets, they had no further military commitment. The committee objected to the costs of special WASP uniforms, and expressed doubts about the necessity for WASP to undergo the Officer Training Course. It concluded that the WASP experiment was "wasted time and wasted effort."

The Costello Bill, slated to come before the House on June 19, still had the backing of the Military Affairs Committee, but it had also now incurred vociferous opposition.

Despite the publicity, Americans had a lot of news to absorb in 1944, and not everyone knew that a woman could be an Air Force pilot. For many, those women in Santiago blue were an unusual sight. One evening in 1944 two WASP were changing trains in Amarillo, Texas. They were in duty uniforms and carrying B-4 bags. It was growing dark, and they passed in front of a drugstore window. They did not notice a car parked across the street, and a teenage girl named Jerri and her mother sitting in the front seat. The girl had large dark eyes, and she fixed those eyes for a moment on the two WASP, and made an astonished pronouncement: "*Mother . . . They're lettin' women in the Air Force.*"

17

DECISION HEIGHT

On June 6, 1944, a few hours after midnight, airborne troops were dropped behind enemy lines in northwestern France to secure bridges and protect the invading force. At dawn, the English Channel was thick with the greatest armada of warships ever assembled, accompanied in the predawn skies by the thunder of eight thousand Allied pursuits and bombers. As troops landed, Allied bombers hit the Germans, and the pursuits patrolled overhead, fending off counterattacks by Messerschmitts. It was a fierce and protracted battle—within a week, thousands of Allied and Axis troops were wounded or dead. But the Allies had secured the beachhead. And Hap Arnold could feel reasonably confident that, at least on the western front, the tide of war had begun to turn.

A month later, in late July, the Allies had yet to advance much beyond the beaches. The Ninth Air Force was sent in to break German supply lines, and the Allies began massive carpet-bombing. On Friday, August 25, American infantry reached Paris.

Almost two years earlier, after the early Allied successes in the Battles of Midway, Stalingrad and El Alamein, Prime Minister Churchill had cautioned, "Now this is not the end . . . But it is, perhaps, the end of the beginning." And in late summer of 1944 many hoped that the liberation of Paris heralded, at long last, the beginning of the end.

In the States there was relief, and hope. The WASP read in papers and saw newsreels of the liberation of Paris, and they also heard of matters on Capitol Hill, where their situation was gaining the attention of the nation's lawmakers. In mid-June the Costello Bill was debated on the floor of the House. The debate was long and confused, its subjects ranging from the mistreatment of male civilian fliers to the propriety of Hap Arnold's jurisdiction over Civil Service funds. It continued for three days, and could have lasted longer, but several more bills were pending before summer recess, a little more than a week away. On June 21, 1944, the Costello Bill, H.R. 4219, was defeated 188 to 169. Those in Congress who felt that Arnold's experiment was "wasted time and wasted effort" did not have the authority to shut the whole program down. But they could and did see to it that the WASP would retain civilian status for the duration.

The WASP were angry and disappointed, but many took the news philosophically, comforted by the simple fact that if they wouldn't be made Air Force officers, they were still flying Air Force planes. In some ways, little had changed. On August 4, the War Department announced that because WASP still required a knowledge of military subjects, they would continue to take the officers' training course at the Army Air Force Strategic Command School in Orlando. Classes were still graduating from Avenger Field, and WASP with newly minted wings were stationed at airbases across the country, and transitioning to pursuits and twin-engine bombers.

In August a twenty-one-year-old woman named Jean Hixson graduated from Avenger Field with seventy-one other members of the class of 44–6. Upon graduation Jean was stationed at Douglas Army Air Field in Arizona, where for several weeks she flew B-25s as an engineering test pilot. Like most of the WASP she was disappointed at being denied an AAF commission, but in September she had a chance to take an advanced instrument training course. On September 12, she wrote her parents, "Cochran is seeing to it that we get the best training in the world. She says we have a right to it."

And in the fall of 1944, one civil servant, recently denied an

Army Air Force commission, would fly an aircraft so secret that few in the Army Air Force knew it even existed.

Ann Baumgartner was a dancer, and when her classmates in 43–5 saw her fly they said it was as though the dancer's precision and timing and grace went from her arms and legs into the plane itself. In January 1944, WASP Baumgartner had been stationed at Camp Davis, where she had been doing test pilot and tow-target work. Then she received word that Cochran was giving her a different kind of assignment. Engineers at Wright Field in Dayton, Ohio, were experimenting with new equipment—oxygen masks and flight suits designed specifically for women—and they needed a test subject. Ann didn't know exactly what this meant, but it sounded interesting and different.

In 1944 Wright Field was where the Army Air Force was conducting experimental research in rocketry, jet propulsion, navigation, and the machines they called computers. The facility was overrun with engineers, scientists, and pilots from the Allied countries. Everybody was moving fast and everybody was talking fast. The talk would go on after hours in bars outside the base. The designers would sit there with their drinks and draw plans for new aircraft on cocktail napkins. They'd talk about how a few years earlier some English engineers who called themselves the British Interplanetary Society had designed a rocket that could actually carry a small crew to the moon. Wright Field in 1944 was an astonishing place.

In a few weeks, Ann's assignment was over. She told the commanding officer that all things being equal, she'd like very much to remain at Wright Field. As it happened, he said, the operations offices at flight test stations needed knowledgeable help. If she put in her time at ops, and if her work was good enough, there was an outside chance she'd get back on the flight line. For Ann it would be a gamble. The thought of spending the remainder of the war filling out checklists and watching those P-51s zooming overhead with somebody else at their controls was almost unbearable. She decided to take the chance at ops, and it worked. In a matter of weeks she *was* back on the flight line, and it was no ordinary flight line. Her fellow flyers were the elite of America's test pilots. They were flying new versions of the P-51 and P-57. There were new pressurization systems, new oxygen equipment, new gunsights. A lot of it was risky. Sometimes she'd be shown a plane with an engine that had gone from the drawing board to the runway in six months. Often enough, engines simply failed, and a pilot would have to land the plane like a glider. A pilot Ann knew was unable to pull out of a steep dive. She saw it happen. He was nearer and nearer the ground and she thought *pull up pull up pull up pull up* and then saw him hit. A moment later she heard the unbearable sound of it, like a crack of thunder a long ways off. Then the smoke in the distance, and the sirens. She could run the whole thing over and over again in her mind like a film loop.

Accidents and death were as much a part of test piloting as they were of combat. But Ann tried not to think about that. At Wright Field there was enough going on to keep her mind off it. For one thing, there were some amazing planes here. Every now and then an Allied pilot would manage to steal an enemy aircraft so that test pilots at Wright could take it up and study it. Wright Field was one of the places these stolen fruit showed up. There was a Mitsubishi Zero and a Messerschmidt 109. Test pilots would fly simulated dogfights against American aircraft. They would determine the plane's maximum speed in climbs, dives and level flight. They would evaluate handling qualities, maneuverability, and engine performance. Their conclusions would then be written up and distributed to combat squadrons. If the simulation had shown, for instance, that a certain German pursuit could climb faster than a P-51, the test pilot might recommend that a P-51 pilot so engaged would be prudent to dive instead.

The engineers at Wright Field were experimenting with something called jet propulsion. In theory it was simple. The principle was known to the ancient Greeks, and in fact the Germans already had a few jets in combat, but the American engineers were encountering some difficulties. In April 1941 Hap Arnold had been astonished to learn that the British government had something called the Whittle jet engine in the final test phase. He arranged an exchange agreement for mutual research and development, and a few months later, in September, an Air Force major flew from London to the United States with a briefcase of drawings handcuffed to his wrist. General Electric was given a contract to develop an American engine based on those drawings, and Bell Aircraft was awarded the contract for the aircraft itself. They worked fast. Three prototypes were ready a year later, and much about it—like its top speed—was classified. When Bell did test flights at Muroc Air Base in California, they were so concerned about spies that they covered the plane with a tarp when they towed it out of the hangar. They had a propeller cut out of plywood with a hole in its hub that fit right over the jet's nose. From a distance, the jet looked like an ordinary propeller plane covered with a tarp. It was called the XP-59A.

In March 1942, Bell had a contract to deliver thirteen service test versions of the aircraft. These were more like actual operational aircraft, equipped with armament, either cannon or machine guns. They were called YP-59A Airacomets. Even as late as 1944 very few pilots knew the jets existed, and only a handful had actually seen one. But there were several at Wright Field. Sometimes the engines could be heard off in the distance. It wasn't the steady, low thunder of pistons. The sound was a high-pitched whine, and there was something unearthly about it.

When Ann did her first walkaround of the YP-59A it seemed sleek but, without a propeller on its nose, strangely naked. And for a pursuit, it looked light. It didn't have the huge engine of something like the P-51. From certain angles it might be mistaken for a glider, except that beneath each wing, flush against the fuselage, were enormous engine housings, huge intake scoops. The air around the plane had the sharp tang of raw kerosene—jet fuel. This was a prototype, and as such it carried only enough of that fuel for three quarters of an hour. The tests would have to be fast.

Ann closed and locked the canopy. Then she pushed the throttle forward and began to taxi. It felt strange to know she was feeding fuel not to propellers but to burners. Like most prototypes it was underpowered, and the take-off run would have to be longer. There was almost no vibration, and almost no sound. Then she gave it power and pulled back on the stick . . . the ground fell away . . . it rose smoothly . . . she climbed steeply. At 35,000 feet she leveled off. Then, with the engines only a high-pitched whine, she tested the pressurization, and fired a few rounds of the .50-caliber machine guns in the nose. She banked a little. Then she tried a slow roll. Even when Ann throttled back, the Airacomet slid through the air like a warm knife through butter. The three quarters of an hour was over too soon, and Ann brought it in. The first test flight of a jet aircraft by an American woman had been without incident. It was September 1944. And for anybody keeping score, the Women's Airforce Service Pilots had now flown every plane in the inventory of the Army Air Force.

On September 20, the ten thousandth P-47 Thunderbolt rolled off the assembly line at the Republic factory on Long Island. This was a cause for celebration, and Republic asked Jackie Cochran to come and break a bottle of champagne across the side of the plane. The Movietone people filmed her doing it, and all the assembled factory workers whooped and cheered. The WASP at New Castle had regularly flown P-47s from Republic to docks in Newark, where they would be loaded onto transports and shipped to the Eighth Air Force. Ordinarily they took their turns in order, but for this one they drew straws. Teresa James won, and so she too was filmed, as she taxied the pursuit affectionately called "The Jug" down the factory runway and roared into the sky.

Jackie smiled for the cameras that day, but privately she was angry. She believed that the defeat of the Costello Bill was a mistake. Period. In August she presented Arnold with an eleven-page report recounting the WASP's achievements. The recommendation of that report was a double-or-nothing gamble: If the AAF would not militarize the WASP, then perhaps the women's corps should be disbanded. In fact, there were already many reasons to end the program. By September, much of Europe had been liberated, and even in the Pacific, the Marines had captured the Marianas Islands, which the Air Force would use as a base from which B-29s could reach Japan. There was no reason to militarize one thousand pilots who would soon be unnecessary. Hap Arnold was ready to cut back the Army Air Force, and Jackie had given him a place to start. The choice was of a kind familiar to pilots everywhere. There is a specific altitude at which a pilot must decide to land, or declare a missed approach and continue flying. It is called "decision height."

In late September 1944, a letter was drafted at AAF Headquarters. On October 3, every WASP received one. Nine hundred WASP—in Long Beach, Dallas, New Castle, Sioux Falls, San Antonio, Colorado Springs, and Greenwood, Mississippi, on more than one hundred Army Air bases ranging from Delaware to Washington, opened the letter and read the same words: "I want you to know that I appreciate your war service and that the AAF will miss you. I also know that you will join us in being thankful that our combat losses have proved to be much lower than anticipated, even though it means the inactivation of the WASP . . . Happy Landings always."* It was signed by Hap Arnold. There was an attachment from Cochran: "To all WASP: General Arnold has directed that the WASP program be deactivated on 20 December 1944."

The women, for the most part, were devastated. They were denied flying, and they would soon be denied full participation in a victory which seemed imminent. On many airbases the mood was bittersweet. Some base commanders allowed WASP to check out in every plane on the field, and the women, knowing it would probably be their last chance to fly a twin-engine bomber or a pursuit, leapt at the offer. Of course, even in the fall of 1944 there was ferrying left to do—for the WASP, exactly eleven weeks of it.

The Lockheed P-38 Lightning looked different from other single-pilot planes. It had a short central fuselage which held the cockpit, and two teardrop-shaped engine cowlings housing its twin Allison V-12 liquid-cooled engines, one on either side. Anyone seeing a P-38 from above might imagine that some sculptor had carved a silver canoe with two silver outriggers. For all their beauty, the planes were outstanding long range fighters, built to fly through flak and survive. They had a maximum horizontal speed of 437 miles per hour, and in high speed dives they had been known to exceed five hundred.† But for the first few years of production the

*Arnold had predicted that an air war would cost fewer lives. The final figures proved him right. The Army Air Force suffered 52,000 killed in action and 63,000 wounded, about 12 percent of the Army's 936,000 casualties.

†Although the fact was not yet widely known, in April 1943 two American pilots in P-38s

P-38 had peculiarities. For one, there was a severe "compressibility" problem during those high-speed dives. It was worse than it sounded. In 1941 a Lockheed test pilot performing a dive had lost control and crashed. It was later determined the plane had been approaching the "transonic zone." The investigation revealed that the P-38 had approached a speed so great that the upper part of the wing, relative to the air around it, had actually been moving faster than sound, but the rest of the wing—indeed the rest of the airplane—was not. The different speeds of air around the wings had created shock waves, and right behind them a turbulence so ferocious that the vertical stabilizers were torn out by their roots.

The P-38 had another peculiarity. As the women who flew with Evelyn Sharp had learned, the nose wheel could stick. Lockheed addressed this difficulty in three ways. It installed two reflectors—small mirrors attached to the underside of each engine housing—that the pilot could look at to see whether the nose wheel was down. It also installed a manual pump with which the pilot could build up enough pressure to force the gear down. Finally, it installed in the wheel housing a CO_2 cartridge which could be exploded, forcing the wheel out. This action, because it could damage the gear, was recommended only as a last resort.

Although the women ferrying aircraft for the Army Air Force had received news of their imminent deactivation, most were still ferrying on a regular basis. In November Nancy Batson got orders to deliver a P-38 from California to New Jersey. It was a long delivery, and as she began the final leg, taking off from an airbase in Pittsburgh, she was looking forward to some rest. But when she was finally in clear sky above the smoke of the steelworks and headed east, she noticed that the needle on the engine coolant gauge was oscillating. She called the tower back at Pittsburgh and requested a return. When she was cleared she made the turn and began her descent. At 2,000 feet she pulled the handle to lower the landing gear. On her instrument panel the red "gear in progress" light came on, and she could hear the hum of the landing gear extending. The whole process should have taken less than a minute, but after two minutes the light was still on, so she looked

---

had found and shot down a Mitsubishi bomber, killing all those aboard, including Admiral Yamamoto, commander of the Japanese fleet.

to the reflector on her left and to the reflector on her right—and saw the nose wheel halfway extended, like a clock hand pointing to four when it should have been pointing to six. She circled the airfield and began to work the manual pump, but it was doing nothing . . . the maddening little red light was still lit and in the right reflector the nose wheel was still pointing at four o'clock.

The tower advised her to try to force the wheel out by centrifugal force, the way one might open a switchblade by snapping the handle forward and then pulling back. He suggested that Nancy try to imitate that motion with the entire P-38. So she took the plane on a long climb out high over the city. At 20,000 feet she pushed forward on the stick, plunging the plane into a steep dive. Then she pushed the throttle forward until her airspeed increased to 350 . . . 380 . . . 410. The altimeter was unwinding, and the steelworks below were getting closer and closer. The twin engines were deafening and the plane was shaking, going faster than she had ever flown it. At 1,000 feet she pulled back hard on the stick. The g-forces pushed her back into the seat and pushed her head back into the crash pad. When she leveled out and the g-forces relaxed she took a deep breath and looked at the light. It was still on. She looked at the reflector and the wheel was still stuck at four o'clock. So again she climbed to 20,000 feet, and again she dived almost straight down and then pulled up. And again she looked at the right reflector and saw . . . the damn idiot wheel still stuck at four o'clock.

Before long the screaming dive bomber high above the streets of Pittsburgh had attracted attention—people stopped to stand on corners and watch. At the airbase the Red Cross and crash crews were standing by. The radioman in the control tower was trying to help her, but there wasn't much he could do or say, and so he said, "How you doing?" And through the static in his earphones he heard her Alabama drawl, "Ah'm still flyin' ovah Pittsburgh and ah'm still pumpin' and ah still have a red light."

Every radio in every plane in the vicinity was tuned to the frequency and every pilot heard her. The local ATC officer joined the others in the tower. Accidents in military aircraft required a protocol. For one, the commander or supervisor directly responsible for the equipment or personnel involved was supposed to be notified promptly. The ATC officer knew he was acting prematurely, but he felt that Nancy Love should know that one of her girls was in trouble, and he managed to get a call through to her in Cincinnati. And through the static of the long distance line the ATC officer heard that whiskey contralto, "Don't you people worry too much. Batson will do just fine."

In the cockpit high over Pittsburgh, pilot Batson was not prepared to pancake the plane or skid it on its belly and risk destroying government property. Instead, she thought she'd give it another try. But by now there was a second problem. Although the needle on the engine coolant had steadied, she'd been diving and climbing for almost two hours and her fuel gauge was reading in the red. If she were to land at all, she'd have to land soon. She took a deep breath and decided to do it all at once. Once more she dove long, and at the bottom, where the g-forces pressed her into her seat, she pulled back hard on the stick and at the same time fired the canister, which made a sharp crack. She thought for a moment she'd blown the wheel clear out of its housing and imagined the whole wheel assembly tumbling down through the air toward the steelworks, but as she started to level off, the light went out and she looked at the reflector and saw . . . a single handsome nose wheel locked perfectly into landing position.

When she landed it seemed as if half of the city was there. As she taxied on the runway she heard sirens—there were ambulances, firetrucks and jeeps racing toward her—she did not want to cause a traffic accident on the runway, so she rolled the plane to a stop. A jeep pulled up behind and an officer jumped out onto the section of wing between an engine housing and the cockpit. In order to hear him, she unlocked and pushed open the center section of the canopy. The officer, thinking he was relieving Nancy of a chore, maybe helping to comfort a frightened little girl, shouted over the engines "Get out. I'll take it in."

But Nancy's assignment was to deliver the aircraft. Although this was a return it was still a delivery, and as far as she was concerned the plane wasn't delivered and her work wasn't over until the thing was parked. She looked at him, shook her head and shouted back, "*Ah've got it this fah. . . . Ah believe ah'll take it home.*"

18

"BUT MY MOTHER FLEW B-17S"

The eighteenth and last class graduated from Avenger Field on December 7, 1944, exactly three years after Pearl Harbor. By that time 1,074 Women's Airforce Service Pilots had earned their wings, and 916 were still on active duty. All told, they had flown over sixty million miles in all weather and in every aircraft in the inventory of the Army Air Force. Thirty-eight had died. More than 100 WASP on active duty found their way to Sweetwater for that last graduation—"returning to the hive," they liked to call it. By then the WAC and WAVES were part of the armed services, with military recognition and the education benefits and insurance of the GI Bill. But it had become clear by then that the WASP were not about to get military recognition, and in fact many would not be allowed to fly for the Army Air Force again. It seemed that on that cold Texas day everyone was admitting that the stepchild was, after all, only a stepchild.

By early afternoon there were B-17s and B-24s sitting on the flight line at Avenger Field—planes in which distinguished guests had arrived. The new gymnasium was set up with hundreds of folding chairs in long straight rows. Some of the girls had made a great replica of WASP wings, fifteen feet wide, and mounted it on the wall behind the stage. The wings for each class had been a little different, and these had at their center the diamond-shaped shield said to be carried by Athena, the goddess of war and protector of the brave and valorous.

In a section for returning WASP, more than one hundred women found seats. Former baymates who were now stationed at bases on opposite sides of the country, and had not seen each other in months, had quiet reunions. The sixty-eight members of the graduating class walked into the gymnasium, a little nervous. Soon the great room was full of young women in Santiago blue. The doors on the side of the stage opened, and there was applause as Jackie Cochran and Hap Arnold and Barton Yount walked onto the stage and sat on the raised dais. General Yount had served as commanding general of the Flying Training Command responsible for all Army Air Force training. He spoke of the thirty-eight WASP who gave their lives saying, "They have demonstrated a courage . . . sustained not by the fevers of combat but by the steady heartbeat of faith."

When Hap Arnold was introduced he stood, walked behind the podium and smiled. He was reading his speech from notes, but his tone was warm, as though he were talking to a good friend. "Frankly, I didn't know in 1941 whether a slip of a young girl could fight the controls of a B-17 in the heavy weather they would naturally encounter in operational flying. Well, now in 1944, more than two years since the WASP first started flying with the air forces, we can come to only one conclusion: The entire operation has been a success." He looked up from the page and his smile grew. "It is on the record that women can fly as well as men."

Then Jackie stood—she was wearing a print dress—and thanked them and said that if the Army Air Force needed them back, she knew that most would be there. She spoke simply and directly. "I am proud that the WASP have merited praise from General Arnold and General Yount. They think the WASP have done a good job. That makes me happy."

And the sixty-eight graduates, wearing Santiago blue for the only time they would ever wear it in an official capacity, sang a song. Each member of the class stood and, one at a time, walked across the stage and stood before General Arnold. Mrs. Deaton held the box of wings, and Arnold pinned them over each graduate's left pocket.

Afterward, inside one of the barracks, somebody said, "I knew it was too good to last," and somebody else said, "Cheer up, kid. Nothing good ever does." A few took sips from a bottle of whiskey somebody had hidden under her bunk. They passed it around and made toasts to the boys overseas and then to themselves. They knew there were still two weeks left, and a few would have a chance to ferry, but that night they felt plain tired.

In the weeks that followed, WASP assigned as ferrying pilots called Nancy Love in Cincinnati at all hours, insisting "They can't do this to me." But she had nothing to tell them. Some WASP tried as best they could to stay flying. Betty Gillies and the forty-two women pilots at New Castle offered to work for the AAF for one dollar a year, but were politely refused. At one base a commanding officer said he disagreed with the decision, and although there was nothing he could do about it, he could honor the WASP on base with a pass and review, a ceremony usually reserved for visiting dignitaries. At one base, WASP received letters from an airline, and it seemed cause for hope. Until they opened the envelopes and read the letters. They had been invited to become stewardesses.

It was Christmastime, and it all seemed so wrong. There was a sense that they were leaving a job unfinished, and a vague and uncomfortable feeling that they were being punished. Teresa James returned to work in her family's flower shop. On December 19, Nancy Batson flew her last P-47 Thunderbolt from the Republic factory out across past the Statue of Liberty and on to Newark. Then she returned to Birmingham. Del Scharr returned to her husband in St. Louis, and before long, to her career as a schoolteacher. Ann Baumgartner managed to remain a Civil Service employee at Wright-Patterson, where she would write publicity about developments in jet propulsion and aerospace research. Some found other ways to serve. Mary Parker volunteered for the Red Cross, and spent the next nine months serving at rest and recuperation centers in the Pacific. Dedie Deaton, who remained in her position until March 15, prepared to write a history of training and administration at Sweetwater.* Nancy Love, as a representative of the three hundred ferry pilots she had commanded, was interviewed by an AP reporter. She said, "I have no plans. I want to get away, and think what all has happened to me . . . and to us."

It was, of course, an eventful spring. On April 12, President Roosevelt died, and Harry Truman assumed the presidency. The war in the Pacific drew to a close that, more and more, seemed inevitable. The Twentieth Air Force began bombing raids on the Japanese mainland. Then, on August 6, the crew of the B-29, *Enola Gay*, commanded by Lieutenant Colonel Paul Tibbetts, dropped an atomic bomb on Hiroshima. By September 2, it was official. The war was over.

In the months and years that followed, the women who had been WASP and WAFS married and had children and thought about other things. Although many would think of the war years as the most exciting of their lives, on the whole, they were women who did not indulge in backward glances. From time to time, a daughter or a son would find the photo album full of pictures—perhaps a photo of three women in oversize coveralls standing beside a PT-19, another of four laughing women awkwardly holding the legs and arms of a fifth and about to toss her over a little stone wall into a pool of water, another of a long row of silver planes. The daughter or son would ask about the pictures. And their mother would tell of practice-strafing an artillery emplacement at Wrightsville Beach, of cruising high above the snowcapped Rockies, and of seeing the New Mexico desert on a moonlit night from the cockpit of a B-26. And later, when someone asked these daughters and sons the famous question, "What did your father do in the war?" they would say something like, "I think he had a desk job somewhere. But my mother? *My mother flew B-17s.*"

The majority of Americans welcomed the return to normalcy

*Mrs. Deaton kept the records in her home in Wichita Falls, until her declining health forced her to store them at the headquarters of the local Red Cross. They have since been lost.

offered by peacetime, and as for women flying those big, dangerous airplanes, why would they *want* to? After a while it just seemed less and less possible, and as the nation looked ahead to the second half of the century, memories of the WAFS and the WASP were swept aside.

In 1946 Nancy Love was awarded the Air Medal, and her husband, Bob, in the same ceremony, was presented with the Distinguished Service Medal. In the following year Nancy gave birth to a girl, and Bob founded Allegheny Airlines. In the fifties they set up residence on the Massachusetts island of Martha's Vineyard. They had two more children, also girls. The family sailed their boat in waters off the New England coast, rode horses on the trails around the island, and took long walks on the shore. Although Nancy had retired from public life, she kept in touch with many of the women under her command, and welcomed them as houseguests.

The women who had been WASP began reunions in years immediately following the war, but attendance was poor. They had new lives now, and the memories could be painful. For almost thirty years there was almost nothing about the WASP in national news. Then, in 1976, a few WASP made a serious effort to locate their counterparts, and a reunion was held in the Velda Rose Tower Hotel in Hot Springs, Arkansas. On that rainy night there were tearful embraces and laughter, and they brought out the scrapbooks and photo albums. For the first time there was also talk of taking action.

It had been thirty-four years since Nancy Love had told a few excited women that they would probably be made second lieutenants in the AAF, thirty-three years since Hap Arnold and Jackie Cochran had studied the possibilities of militarization. And now the talk began again. A few weeks before, Senator Barry Goldwater of Arizona, himself a former pilot with the ATC in New Castle, had introduced an amendment granting veteran status to the WASP. He asked that it be attached to an obscure bill which provided Veterans' Administration medical benefits to naturalized Poles and Czechs who had fought as Allied soldiers. The amendment passed the Senate, but a few days later the House rejected it. No wartime civilians had ever been accorded veteran status, and veterans groups believed that if it passed, the Merchant Marine,

Civil Air Patrol, and a half dozen other organizations would think themselves entitled to the very finite resources of the veterans affairs budget. But the door had been opened at least a little. The House and the Senate Veterans Affairs committees promised that the next Congress would decide whether to bestow veteran status on the WASP.

As 1977 began, Senator Goldwater introduced a bill "to provide recognition to the Women's Airforce Service Pilots for service to their country by deeming such service to have been active duty in the Armed Forces of the United States for purposes of laws administered by the Veterans' Administration." W. Bruce Arnold, son of Hap Arnold and now a retired Air Force Colonel, got involved in the national lobbying efforts. He told the WASP that he was finishing the job his father had begun. Congresswoman Lindy Boggs sponsored the bill in the House. The House committee's only woman was Margaret Heckler of Massachusetts, a member with a distinguished history as a spokesperson for women's rights. Through her efforts in the subsequent weeks, every congresswoman cosponsored the bill. Meanwhile, the WASP themselves, among them Teresa James, Byrd Granger and Nancy Batson, began collecting signatures on petitions to present to veterans' organizations. Soon much of the national media was behind the effort, and even *Stars and Stripes* lent its support. Then, on May 25, the Senate Veterans Affairs Committee began hearings.

The WASP had become used to their service being unrecognized, and it did not surprise them that it would be forgotten. Then something happened which seemed to say that the program had never even existed. On August 9, the Air Force issued a press release announcing that "*for the first time the Air Force is allowing women to fly its aircraft, saying 10 women did as well in training schools as successful males.*" WASP read this, and then read it again. *For the first time*? In fact, those ten women caused the appearance of a surprise witness in the hearings of the House Veterans Affairs Committee on September 20. Antonia Handler Chayes, an assistant secretary of the Air Force, was concerned about the effect a defeat of the bill would have on the morale of those women, and subsequent recruitment efforts. Also testifying on behalf of the WASP was commanding officer of the Ferrying Division, William

Jacqueline Cochran, director of the Women Airforce Service Pilots, wearing the WASP uniform *(Courtesy of the International Women's Air and Space Museum)*

Nancy Harkness Love, founder of the Women's Auxiliary Ferrying Squadron, wearing an A 2 flight jacket with a Ferry Division patch *(Courtesy of the International Women's Air and Space Museum)*

Henry "Hap" Arnold, commanding general of the Army Air Forces, in 1944 *(Courtesy of the National Air and Space Museum)*

Instructor demonstrating a slow roll in hanger number 3 *(Courtesy of the Texas Woman's University Library)*

Trainee Marjorie Harper filling out Form 1-A *(Courtesy of the Texas Woman's University Library)*

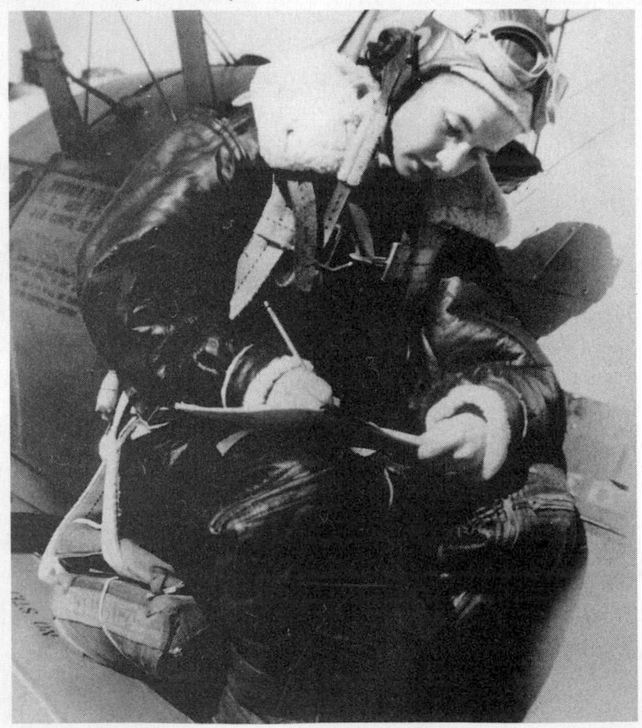

Trainees being briefed in the ready-room, Avenger Field, Sweetwater, Texas, May 1943. Hazel Ying Lee is second from the right. *(Courtesy of the Texas Woman's University Library)*

Wilda Winfield in flight in a BT-13 from Frederick Air Force Base, Oklahoma *(Courtesy of the Texas Woman's University Library)*

The trainees from the class of 43-W-6, celebrating a successful solo *(Courtesy of the Texas Woman's University Library)*

Trainees Jo Martin, Elizabeth Ann Lore, and Clara Jo Marsh walking along the line of AT-6s at Avenger Field after an unusually heavy snowfall *(Courtesy of the Texas Woman's University Library)*

Jane Straughn carrying a parachute and a B-4 bag after delivering a pursuit to Ellington Field, Houston, Texas *(Courtesy of the Texas Woman's University Library)*

WASP reviewing the flight plan on the wing of an AT-6. Nancy Batson is second from the left. *(Courtesy of the Texas Woman's University Library)*

Jacqueline Cochran meets with the WASP at Camp Davis, North Carolina *(Courtesy of the Texas Woman's University Library)*

Betty J. Hanson, after delivering a medium-weight bomber at Newark, New Jersey, airbase *(Courtesy of the Texas Woman's University Library)*

The last WASP class graduation, 44-W-10, Avenger Field, Sweetwater, Texas, December 1944 *(Courtesy of the Texas Woman's University Library)*

Jean Hixson and Jackie Cochran holding a model of a North American F-86 in 1955 *(Courtesy of the International Women's Air and Space Museum)*

Jerrie Cobb at the Tyndall Air Force Base in 1959, moments before riding backseat in a Convair TF 102 "Delta Dagger." She would act as pilot briefly at 42,000 feet, although she would not be allowed to take off or land. *(Courtesy of the NASA History Office)*

Jerrie Cobb and the Mercury capsule mock-up *(Courtesy of the NASA History Office)*

Jan and Marion Deitrich in 1961 *(Courtesy of the International Women's Air and Space Museum)*

Gene Nora Jessen, Wally Funk, Jerrie Cobb, Jerri (Sloan) Truhill, Sarah (Gorelick) Ratley, Myrtle Cagle, and Bernice Steadman at Cape Canaveral for the launch of STS-63 *(Courtesy of the NASA History Office)*

Tunner. Although both witnesses had well documented evidence, the opposition, mostly veterans' groups, was articulate and sure—the status of veteran had a long and hallowed history in America, and to grant that status to these women would denigrate it. The leadership of the committee did not regard the conditions under which the WASP served as comparable to those of established veterans. Then, on October 19, Senator Goldwater selected H.R. 8701, the "GI Bill Improvement Act," to host his original amendment. The legislation involved education benefits, and both Veterans Affairs Committees wanted it passed. In a matter of days the leadership of the House Veterans Affairs Committee announced that if the Air Force would certify that the WASP had been de facto military personnel, they would support it.

Meanwhile, Bruce Arnold had been compiling documents from former members of the WASP that might demonstrate that they were military personnel. He found formal discharge papers of WASP Helen Porter, which certified that she had served honorably in "the Army of the United States." WASP discharge papers were of various types, and not all WASP received discharges, yet clearly someone in an official capacity in 1944 had considered Helen Porter to be Army personnel. It was only a single piece of paper, but it was enough, especially for Congressman Olin E. Teague. Teague was chairman of the subcommittee, and a former colonel in the Army. Helen Porter's discharge was identical to his own.

On the afternoon of November 3, the House passed the bill unanimously. The following evening the Senate approved it. Bruce Arnold, and a few WASP in Santiago blue were watching from the visitors gallery. When the gavel struck, Arnold noticed a tear rolling down the cheek of the woman sitting beside him. He pressed her hand in his, and whispered, "Stop that. You're in uniform."

The gathering of women for political purposes had an unintended side effect. At Sweetwater now, there is a plaque listing the name of every WASP to earn wings, and near the names of the thirty-eight who were killed, there are small gold stars. In the center of the Wishing Well—the same wishing well in which Lorainne Zillner and every trainee landed after her solo—there is a pedestal and a small statue of a young woman in a flight suit and a leather helmet. The WASP bring their daughters and their granddaughters and even their great-granddaughters to the reunions, and they toss a coin into the well and look at the statue of the young woman. She is looking upward.

So for a few months in 1977, the passage of H.R. 8701 put the WASP in the national news again, but the attention would be short-lived. By then too much water had gone under the bridge. Names like Jackie Cochran and Hap Arnold and Nancy Love had been forgotten. And the planes—the machines they feared and hated and loved, the machines which sustained them and which killed them—had long since been put in museums or dismantled for scrap. Today, if you ask people on the street who the WASP were, some of them might say they were secretaries who worked for the Air Force in World War II. Others might tell you they were a women's volleyball team.

And most would probably say they had no idea.

PART II

19

NOBODY'S ANGELS

In the months and years after the WASP were disbanded, American women found it difficult to pursue careers in aviation. The Army Air Force had produced tens of thousands of new pilots, many of whom returned to civilian life with veteran status, a few thousand hours in twin-engines, and a determination to continue flying. Competition for civilian piloting jobs was fierce. Across America there was much talk of a "return to normalcy," meaning a husband who worked and a wife who attended to the household and the children. The working women a year earlier regarded as patriots—the Rosie the Riveters and the Wendy the Welders—had suddenly become a social problem.* Similarly, a woman who wanted a career as a pilot was regarded as a threat to the livelihoods of male pilots, many of them veterans.

*By 1943 female employment in aircraft engine plants had been 31 percent, and one industry spokesman estimated that women performed half the work necessary to construct an airplane. By 1946 three million women had left work, a great many of them laid off. Few employers expected they would want to continue, but many demanded reinstatement. Although unions supported women's protests, few would press the matter, and as the war plants converted to peacetime operation, men, especially veterans, were generally hired over women.

After the war there was no ready means for a woman to pay for flying lessons, or even if she had earned a license, to lease a plane. It was true that a girl could join the local chapter of the Civil Air Patrol, and might be introduced to an instructor who would give flying lessons at a discount. In previous decades, aspiring women pilots either had a means to generate income, barnstorming in the 1920s, for instance, or enjoyed the benefits of a government program like the CPTP and the WASP in the 1940s. But in 1950 the public was not interested in barnstormers, and the WASP had been disbanded.

So for American women things weren't a whole lot different after the war than they had been before it. But aviation itself began to change radically. It had entered the jet age. The Air Force was developing the F-84 and F-86 fighters and the B-52 bomber. In 1946 a Lockheed F-80 Shooting Star crossed the continent in four hours and thirteen minutes. The next year Captain Charles "Chuck" Yeager flew Bell Aircraft's X-1 faster than the speed of sound. In the Korean War, F-86s would be used in aerial dogfights. The men given the opportunity to fly all these planes were trained at government expense, in Air Force or Navy training schools, institutions which did not allow entry to women. For the first time in aviation history, it was impossible for a woman pilot to aspire to a place at the forefront of her profession. The forefront of the profession belonged to jet pilots. And women aviators in America were limited to the propeller age.

Still, there were women in America who continued to learn to fly, and to look for work as pilots, mostly because they simply couldn't imagine doing anything else. It wasn't easy, but it was possible. Long hours of odd jobs after school might pay for a lesson or two on the weekend. And once a woman had earned her license, she might hope to find a post as an instructor, or occasionally pick up a charter flight out of her local airport. She might not make a living at it, but if she was lucky, she would at least earn enough to be able to keep flying.

As in previous years, many women pilots were regarded by their peers at best as rather odd, and at worst as "bad girls" or "Airport Annies." As in the 1920s, a woman pilot was likely to be the only woman at the airport. Since the deactivation of the WASP, American women aviators were in the curious but strangely familiar position of the outcast.

The stories of two women who were aviators in the 1950s are illustrative in this respect. The younger of these was a young Oklahoman named Jerrie Cobb. Jerrie first flew in 1943; over the next decade or so her experience as a pilot was a constant struggle to fund her aspirations, requiring long and often fruitless quests for employment. Hers was fairly typical of the experiences of American women pilots. The experience of the second woman was anything but typical. She enjoyed enormous financial resources, as well as political and personal connections with the powerful. Before the war, she had been one of the hottest pilots, male or female, in the world. After the war, Jackie Cochran would become one of the trailblazers of the jet age.

1943

Jerrie Cobb was born in Oklahoma in 1931. But by 1943 her family was living in Wichita Falls, Texas, where her father was stationed at Sheppard Army Air Base. She was the younger of two daughters. She and her sister, Carolyn, were close, but they weren't much alike. Carolyn played with dolls; Jerrie rode horses. She had a pony of her own—she'd earned the money for him herself, selling vegetables that she had grown in the backyard. And whenever she could, she liked to ride out past the edge of town, until she could see nothing but mile after mile of rangeland, and the great yawn of the open sky.

When the war began, Jerrie's father was too old to serve in the infantry, so he enlisted in the Army Air Force, moved his family to Texas, and at Sheppard Field learned to fly. He earned a private license easily enough, but to fly for the Air Force he needed a commercial license, and that required 200 hours. He decided that if he was going to have to log that many hours, he might as well buy an airplane of his own.

The plane Harvey Cobb bought was a tiny, secondhand Taylorcraft. It didn't look like much, but he was proud of it, and he wanted his wife and daughters to enjoy it as much as he did. So one day he took each of them up in turn for a ride. Jerrie didn't expect it would compare with horseback riding. On the ground the plane was an awkward little thing; as it taxied down the runway it sputtered and coughed and bounced. So she was utterly unprepared for what happened when the wheels left the ground and the plane began to climb. In the air, the little Taylorcraft was different. It was graceful and smooth, and it soared. Beneath it the vast distances of the plains sprawled out in such a way that it became obvious that even on horseback she had never done more than dance around their edges. She was astonished that such a thing was possible.

When they had landed and taxied to a stop, Harvey Cobb turned to his daughter, grinned and said, "Well, what did you think?" And Jerrie just looked at him, her head reeling a little. And then her father said, "Well—did you like it?" Finally the words came: "I want to learn. Teach me to fly."

But Jerrie's mother also found out what she thought about flying. She found it terrifying. She could not trust a flimsy contraption of metal, fabric and wood to keep her suspended in the air. When Jerrie said she wanted to take flying lessons, her mother just said no. Jerrie asked again the next day, and the next, and the next. And finally her father said, "If you can get your mother to agree to it, then I'll teach you." It took a little longer before her mother came around, but at last she just sighed and said okay. The next afternoon, Harvey Cobb gave his daughter her first flying lesson. By then, he had traded in the Taylorcraft for a Waco, a biplane with a maximum air speed of 92 miles per hour. It had a wooden propeller and fabric-covered wings, and it looked like a poor relation of the enormous, sleek aircraft at Sheppard Field. But to Jerrie, that hardly mattered, because for an hour on that afternoon, when her father had let her take hold of the stick and begin to feel how she could bank the plane, first to the left and then to the right, she had actually been a pilot.

This was just the beginning, of course. It would take many more hours of instruction before she could fly solo, and before she could earn a license. But she had plenty of time. She was twelve years old.

1945

When the war was over, demobilization did not happen quickly enough for American soldiers stationed in Manila, who began to stage public protests and send cables to congressmen. In the first months of 1946 there were sympathetic demonstrations on bases all over the world. They were called the "Wanna-Go-Home Riots." Like the soldiers in Manila, most military personnel were more than ready to put the war behind them. Jackie Cochran felt rather differently. Over the last few years she had grown accustomed to the frenetic pace of wartime life, and when the WASP were disbanded she found that she wasn't ready to stop. She needed a period of adjustment to the quieter rhythms of peacetime.

So in 1945 Jackie became a war correspondent for *Liberty* magazine. By her own admission, she had never been much of a writer, but she had connections. Her assignment was to go to Asia and simply report back on what she saw. Few American civilians could get to Asia in the summer of 1945, but Jackie talked her way aboard military transport planes, first to Hawaii and then to Guam. Several of the officers in the occupying force did not want her there—they had heard the Cochran stories even across the ocean—but Jackie had an ace up her sleeve. In addition to her journalistic assignment, she had orders as a special consultant to Hap Arnold. The orders were vague. They said she was supposed to look around in the course of her travels and report back to Arnold on any matters of benefit or interest to the Air Force.

In September, she became the first American woman to set foot in Japan after the bombs were dropped. She toured the devastation of Tokyo and Kyoto with a party of Air Force officers. A few weeks later, she traveled still farther from home, to China, where her connections gained her audiences with both Mao Tse-tung and Madame Chiang Kai-shek. She rode a series of military transports westward across Asia and Eastern Europe, pausing finally in Rome, where her friendship with Cardinal Spellman gained her a visit with the pope. In November she headed north to Germany. The Nuremberg Trials were beginning, and as it happened, they were being headed up by another old friend, U.S. Supreme Court Justice Robert Jackson. A private wire to Jackson gained her a seat, and there she heard several firsthand accounts of the war's final days. Of particular interest to Jackie was Hermann Göring's testimony. He said he knew the Third Reich was over the day he saw Thunderbolts and Mustangs flying over Berlin. Jackie felt some satisfaction in the knowledge that some of those planes had been ferried by her WASP.

The articles for *Liberty* magazine never did get written, but Jackie returned home from her world tour with a souvenir for Floyd—a doorknob from Hitler's underground bunker in the Riechschancellory in Berlin, cadged while the Russian soldiers who were guarding the place politely looked the other way.

Jackie learned that she had been chosen to receive the Distinguished Service Medal (DSM) for her work with the WASP, and that Truman himself would present it to her. She was honored insofar as the DSM was the second-highest honor bestowed by the American military. But she told the president's intermediary that if she deserved the medal, it was because of Hap Arnold. If he couldn't present it, she didn't want it. So arrangements were made. A few weeks later the ceremony took place in a Pentagon conference room. The commanding general of the Army Air Force presented Jacqueline Cochran with the Distinguished Service Medal, and her eyes grew bright with tears.

1946–1947

In 1946 Jackie Cochran went on inactive Air Force duty, and returned to racing and to running her cosmetics company. In 1946 she purchased a P-51 Mustang at an Air Force surplus sale and flew it to a second-place finish in that year's Bendix race. She calculated that she was logging some ninety thousand miles a year in the air, traveling around to department stores all over the country on behalf of Jacqueline Cochran Cosmetics. It was a busy schedule by any standards. And then, Jackie decided to go back to work for Hap Arnold.

Hap Arnold had retired on June 30, 1946, keeping a promise to his wife that they would spend their last years on their farm in the countryside near Sonoma, California. Although he told a friend he intended only to watch his cows, he attended closely to developments in the Air Force, and still made himself available to reporters, who often quoted him criticizing postwar spending cuts. Arnold was particularly outspoken on the subject of Air Force autonomy, a cause which he had espoused since the twenties. Now that the war was over, he felt that the Air Force had clearly earned its independence. He had voiced his frustrations to Jackie Cochran, and Jackie suggested that although he might never persuade the Army itself to cut the Air Force loose, he could pitch his case to a different audience, one that was likely to be more receptive. Convince the American public that the Air Force deserves independence, she said, and the Army will have to listen. She proposed a massive public relations campaign, and she volunteered herself as a spokesperson. She spent half a year stumping around the country on behalf of a separate Air Force. She barraged the newspapers with stories about what the Air Force had done to win the war. She even took out ads in the papers at her own expense.

On July 26, 1947, Congress passed the National Security Act of 1947, which established the Air Force as a separate branch of the armed forces. To a large degree, the Air Force was made autonomous because Arnold had convinced Army Chief of Staff George Marshall that the idea had merit. The degree to which Jackie Cochran's efforts contributed is uncertain, but certainly they helped. Jackie considered her work nothing more than a debt repaid, a thank-you present to Hap Arnold.

The swift rise in Arnold's responsibilities during the war had exacted a toll on his health. Over the previous decade he had suffered four heart attacks. On the morning of January 15, 1950, Hap Arnold died of a fifth. He was sixty-three.

It had been a full and astonishingly productive life. Henry "Hap" Arnold had guided the swift creation of the greatest air force in history, expanding it from a few thousand planes and officers to more than 70,000 aircraft and 2.4 million personnel. He had planned and overseen military air operations that had helped defeat the Axis in Europe and Japan in the Pacific, and had helped end a major war without a land invasion—a historic first. He had advocated and initiated a research and development culture that would define the Air Force long after his death. And he had approved and overseen an innovative social experiment in military aviation called the Women's Airforce Service Pilots.

1949–1950

By the time she was sixteen, Jerrie Cobb was flying solo. A student pilot who had soloed and was at least seventeen years of age was eligible for a private pilot's license. Jerrie Cobb passed the test for hers on March 5, 1948, her seventeenth birthday. Exactly one year later she passed the test for a commercial license, which entitled her to call herself a professional pilot and to hire herself out as such to anyone who wished to make use of her services.

But an eighteen-year-old girl was not most people's idea of a professional pilot, and when Jerrie went looking for work, the responses were less than overwhelming. So she decided that, since she was about to graduate from high school, she would simply find a job of some kind, and let the job finance the flying for a while, until she'd racked up enough hours that the record in her logbook could quell doubts about her youth. She'd always been a pretty good softball player, and professional women's softball was a popular spectator sport, so she tried that first. Her tryout went well, and she was offered a chance to play first base for the Oklahoma Sooner Queens.

But her parents had other ideas. They wanted her to go to college. They appreciated her love of flying, but they were pragmatic. Pilot jobs were scarce. She would not be able to count on employment in aviation. College meant a secure future, and softball was no surrogate. Jerrie viewed college as four years of not having much time to fly, so she protested that she didn't need another degree. Her parents didn't back down. Neither did she. At last, however, they reached a compromise. Jerrie could work—and fly—for a year if she would start college the subsequent year. After a year Jerrie's way and a year her parents' way, they would renegotiate.

Jerrie spent the spring and summer of 1949 playing softball and saving her money, and in the fall she enrolled at Oklahoma College for Women at Chickasaw. Classes were more interesting than they had been in high school, but she was still drawn to the airport. After school and during vacations she worked as a general flunky for other pilots, and eventually she was hired part-time by a crop-dusting service. In late spring, when it was time to renegotiate, Mr. and Mrs. Cobb told Jerrie that they would defer to her wishes.

But real work—full-time work that would pay rent and electric bills—was still unavailable to an eighteen-year-old girl pilot. She signed on with the Sooner Queens for a second season. They were playing a series in Denver when she saw a plane for sale. It was a war surplus Fairchild PT-23, and it cost five hundred dollars. She borrowed exactly that amount from the team, and by the end of the summer, she owned her own airplane. A pilot-for-hire who owned an airplane had something of a leg up on the competition. In the fall of 1950, she was hired part-time by an oil company to patrol its pipelines for leaks. It was monotonous work, flying low over the fields of Oklahoma and Kansas, but it was a start.

In 1950, the Harmon Trophy selection committee pronounced Jackie Cochran the most outstanding female pilot of the preceding decade. The same year, Jackie, in her North American P-51, broke the international speed record for propeller-driven airplanes. By 1950 few pilots anywhere held more aviation records than Jackie Cochran. The list was so long that even Floyd couldn't reel it off anymore.

1952–1953

In 1952, when she was twenty-one, Jerrie Cobb was offered work as a flight and ground school instructor. She was in airplanes with students all day, and talking flying at night. But it wouldn't last. Two and one half months after she started teaching, the school had its certification revoked. Jerrie was young and resilient and—she was careful to admit—fortunate. In a few weeks she found work as a charter pilot out of Downtown Airpark in Oklahoma City. She ferried passengers all over the region, to small- and medium-sized cities in Texas, Kansas, and Colorado. But pilots for the charter service were expected to perform other jobs around the airport as well. Some worked as mechanics, some as ground crew. Jerrie was assigned to wait tables at the airport restaurant. On more than one occasion, she served her passengers lunch before she flew them to their destinations. One customer started requesting the "flying waitress" as his regular pilot.

In 1953 Jackie Cochran was at least forty-three and probably closer to forty-seven. She held dozens of records in propeller planes—for altitude, speed and distance. But some days out at the ranch in California she would look overhead and see a long contrail growing across the sky. She would remark to Floyd that maybe the jet age was passing her by. Floyd knew this rare sadness in her voice, and did not like it. And so they talked. It became clear that Jackie wanted more than to pilot a jet—she wanted to become the first woman to break the sound barrier. For that she would need a suitable course, with systems for tracking and recording speeds and flight paths of supersonic aircraft. She and Floyd both knew of just such a place. It lay in the high desert 120 miles to the northeast.

Rogers Dry Lake is a flat expanse some sixty-five square miles in area. The site was originally settled by Clifford and Eddie Corum in 1910, and the small community that grew on its edge was called Muroc—Corum spelled backward. In September 1933 the Army Air Corps set up a gunnery range on the lake bed, and after the attack on Pearl Harbor they built a base to train bomber pilots. In 1942 it became the test site of the XP-59A. After the war it was called Muroc Air Force Base, and there the newly autonomous Air Force tested experimental aircraft. Some were saying it was the birthplace of the Space Age.

In January 1950 Muroc was renamed Edwards Air Force Base for Captain Glen Edwards, a test pilot who, two years earlier, had been killed flying Northrop's Flying Wing. In 1949 the Soviet Union exploded its first atomic bomb, and the U.S. responded by spending millions of dollars to rebuild the Air Force. By the mid-1950s Edwards had become the second-largest Air Force installation in the country. Four government agencies and ten manufacturers had established permanent organizations there. There was a rocket engine test stand for static testing, a high-speed track for ground tests at speeds up to 1,500 miles per hour, an all-altitude speed course, and a precision bombing range.

Jackie and Floyd knew she would also need a suitable aircraft—a North American F-86. It was an aircraft with an interesting history. After the war, a request by the Air Force for a fighter capable of high speeds was answered by North American Aviation. Soon they had on the drawing boards an aircraft they were calling the F-86 Sabre. The wings of the first models went straight out at ninety degrees to the fuselage, but wind tunnel tests were showing that a plane of that design could move no faster than about 600 mph. Someone observed that certain documents recovered from the Germans had shown that a swept-wing design could move faster without an increase in engine thrust. They tried it, and it worked. By 1947 North American had at Muroc three F-86 prototypes thought capable of attaining supersonic speeds. And on April 26, 1948, barely a year and a half after Chuck Yeager had broken the sound barrier in the X-1, a test pilot named "Wheaties" Welch had performed the same feat in an F-86.

But an F-86 Sabre was a million-dollar high-performance jet, and every one of them belonged to the Air Force. As it happened, however, a Canadian aircraft manufacturer called Canadair, under license from North American, was producing its own F-86. Because it was fitted with its own Canadian-designed Orenda engine, the Canadian F-86 was a little more powerful than the American version.

But if Jackie were to break records, she needed more than fast planes. She needed Edwards Air Force Base, which took some lobbying. Exactly who she and Floyd spoke with is unclear, but certainly the list included Joint Chief of Staff Curtis LeMay, Secretary of the Air Force Stuart Symington, Edwards base commander General Albert G. Boyd, Edwards director of flight testing General Fred Ascani, and General Hoyt Vandenberg. Her request was irregular, but permission was granted.

Finally, she needed instruction. Here at least, there was a simple and ready answer. She had met Chuck Yeager in Washington in 1947, shortly after he'd broken the sound barrier. She pumped his hand and said, "Great job, Captain Yeager. We're all proud of you." And then she said, "I'm Jackie Cochran." She said it as if he could just fill in the rest, when in fact at that particular moment Yeager had no idea who she was. But he would learn fast. She, Floyd, Chuck, and his wife, Glennis, would become best of friends, and Yeager would learn to admire and care for Jackie, and be exasperated by her in turn. There were no doubts about Yeager's qualifications as a flight instructor for supersonic aircraft. Chuck Yeager was probably the best pilot in the world.

When the training started, Jackie would have to be ready each morning at 5:00 A.M. It seemed impractical to make the trip from the ranch all the way up to Edwards every day, so she and one of her housekeepers moved up into the Yeagers' modest home, while Glennis Yeager and her four children moved into the ranch with Floyd. Chuck Yeager himself bunked at Edwards' bachelor officer quarters. At Edwards that spring, there were a lot of jokes about Cochran. Colonel Fred Ascani was director of flight testing. He had 120 test pilots working for him, but he liked to say that he had seven prima donnas under his wing, meaning his best test pilots. He began to call Cochran number eight.

The F-86 in 1953 wasn't nearly as dangerous as the X-1 had been in 1947. But there was one significant difference. The X-1 engine was so powerful that it could go supersonic in horizontal flight. If something went wrong in midair, the pilot at least had altitude, which provided some time to recover. The F-86, however, was not as powerful. It was a jet, not a rocket-plane, and to gain supersonic speeds it required an assist from gravity. To break through the sonic wall, the pilot had to go into a steep dive, actually pointing the aircraft at the earth. Then, to arrest the dive, the plane had to be pulled up sharply, so sharply that the pilot encountered g-forces strong enough to cause a great deal of physical pain.*

During the war, it became apparent that pilot blackout, which occurred during dives and turns, was caused by loss of blood in the upper part of the body—particularly the brain. So specialists in aviation medicine had developed what was called a g-suit. It put pressure on the waist, and kept the blood from moving to the lower part of the body, preventing blackout. F-86 pilots were supposed to wear g-suits, but not all of them did. With strong stomach muscles and sheer will, an experienced pilot could avoid blackout without one. In part because the other pilots weren't wearing g-suits, Jackie didn't either. Which was not to say that Jackie didn't have

*In fact, in the growing repertoire of test pilot legends, the one told most often concerned this very subject. On the morning Yeager had broken the sound barrier, he was favoring two cracked ribs he had broken falling from a horse two nights before. He had not reported to the base flight surgeon because they were well into testing, and he knew he would be grounded.

her own kinds of preflight procedures. F-86 cockpits had a raw odor of kerosene and sweat, and Jackie hated both, so she wore a lot of perfume, and sometimes she spritzed some around the seat.

With Yeager flying chase in an Air Force jet, Jackie learned how to put the F-86 through its paces. Every morning she went up and took it faster and faster. Finally the day came. Jackie in her F-86 and Yeager in another climbed to 40,000 feet, then turned their noses straight down. She pushed the throttle forward and watched the airspeed indicator climb upward, and felt the plane buffeting. Yeager had told her to tell him everything she felt and saw, so she reported the instrument readings aloud, saying, "Point eight . . . point nine . . . nine one . . . nine five . . . Mach one . . ." Suddenly the buffeting smoothed out and everything was quiet. She was moving faster than sound. At that moment she felt as though she were inside an explosion, but all she said was, "Shock waves . . . look like rain."

A double sonic boom sounded across Rogers Dry Lake. It was May 18, 1953 and Jackie Cochran had become the first woman to break the sound barrier.

A few minutes later they landed. Jackie climbed out of the plane and walked over to Yeager. She stopped a few feet in front of him, threw her shoulders back as though at attention, and, with exaggerated military solemnity, said, "Thank you, Major Yeager." Then she broke into a smile and hugged him.

By 1952 Jerrie Cobb had logged more than 2,000 flying hours and held commercial and instructor ratings. She had a part-time but steady job as a charter pilot in Oklahoma City, and if she had to double up as a waitress, she knew that just about any pilot who worked out of a small airport was expected to do double duty of some sort. And she was young. In time, as she gained experience, other opportunities would come along.

But in November her father called, and asked for help with his business, a car dealership about one hundred miles north, up in Ponca City. Jerrie quit her job. For a few months things were so hectic that she hardly had time to think about flying, but then work began to ease up and she began job hunting. It was the same story everywhere—so many veteran pilots looking for work that few people even bothered to glance at the résumé of a twenty-one-year-old girl.

Then Jerrie discovered another way to fund the flying. Like Amelia Earhart and Jackie Cochran before her, she took up racing. By the fifties, all-women's races had become a serious business. A woman would persuade an aircraft manufacturer to "sponsor" her—that is, provide her with one of its planes for use in a given race. If she won or placed well, the manufacturer had a new selling point for that plane, and the contestant could show the next sponsor a record of wins or near wins. There was prize money too, not always a lot of it, but enough that a good pilot could at least be assured of covering her flying expenses.

In 1953 Jerrie Cobb entered her first race. The Skylady Derby was a five-hundred-mile course from Dallas to Topeka. She was the youngest contestant, but she placed third. It seemed an auspicious start. A week later, she took off for California to enter the twenty seven hundred-mile All Women's Transcontinental Air Race. It was the most prestigious and perhaps the most difficult women's race in the country, the descendant of the original Women's Air Derby of 1929. In 1953 it was a long and grueling course, from Santa Ana, California, to Teterboro, New Jersey. And Jerrie had her own difficulties. Upon nearing the finish, she was given landing instructions: "TAR three six, Teterboro tower. Look for several tall buildings to the north, also double railroad tracks, converging rivers, and a six-lane highway." To a girl raised in the Midwest, these instructions described most of the East Coast. She was lost for half an hour before she was able to find Teterboro. She still finished fourth.

She began to make something of a name for herself in racing, but she was having no luck finding a steady and full-time job as a pilot. Even with the prize money starting to come in, flying was expensive, and the races were too few and far between to provide a real income.

In the spring of 1953 Jerrie heard from a friend that Miami was "America's gateway for aviation," with plenty of pilot jobs. She moved there, only to discover that she had arrived during the off season, when flying jobs were as scarce as they were anywhere else.

But she scoured the want ads daily, and at last saw something that looked promising. A start-up airline called Trans-International was looking for pilots and stewardesses. Logbook in hand, she found the office in an old hangar and offered her services as a pilot. The two men in the office just looked at her. Then they looked at her papers, and one said, "Well, why not? Maybe a woman pilot is just what this outfit needs. It'll be different anyhow, good publicity." The other one added, "Besides, she can double as a stewardess." She was hired. They were using her as a publicity stunt, and making her go to stewardess school, but she didn't mind because they were also going to let her fly their DC-3s. Once they got those DC-3s. The financing, they explained, was a little sticky at the moment. Jerrie waited, and then the financing fell through altogether.

Eventually, she had no alternative but to seek a desk job. She found one in the customer service department of an aviation firm called Aerodex, Inc., at Miami International Airport. There her luck finally turned.

When she first met him, Jerrie would never have imagined that Jack Ford could offer her a real career in aviation. Their first encounter was inauspicious. She arrived at the office an hour early and found him waiting, an unshaven man who looked to be somewhere in his thirties, grumpy and disheveled, complaining about a paperwork snafu that made him wait for repairs in Miami when he and his plane were needed urgently in South America. He was rude and impatient, but Jerrie filled out his work order anyway. When he left in search of a mechanic, she hoped she'd seen the last of him.

Hours later he was back, showered, shaved and neatly dressed—and in a markedly better mood. Seeing Jerrie at her desk, he smiled and invited her for a cup of coffee—as a way of saying thank you, he explained, for her help in expediting his work order. Her first impression of him still lingering, Jerrie accepted reluctantly. When they were seated in the coffee shop and he began to talk to her about his job, however, he suddenly started to seem like someone she might want to know after all. Jack Ford had flown bombers in the Eighth Air Force during the war, and afterward had begun an aircraft ferrying service called Fleetway. At the moment, he told her, many of his biggest clients were in South America. But he had a problem: It wasn't easy to find experienced pilots who were willing to fly single-engine planes over a long route that was mostly ocean and jungle.

Jerrie didn't hesitate: She laid out her credentials and offered her services. Jack Ford said no. He said flying was a man's job, a tough, risky life, a career that required the kind of solitude and self-reliance that women simply weren't cut out for. Jerrie responded that he might be surprised to know just how many women had proved that they were cut out to be pilots; she reiterated her credentials. He admitted that they were impressive. But his answer was still no; there was no room at Fleetway for a girl pilot.

She went back to her work and forgot about him. A week later she got a phone call. Jack Ford had a whole fleet of T-6s sitting in New Jersey, waiting to go to Peru, with no pilots to fly them. If Jerrie wanted the job, it was hers.

The North American T-6 had been used by the AAF in the war as a single-engine advanced trainer, and in fact there had been a great number at Avenger Field, where they had been called AT-6s. By 1953 the U.S. Air Force no longer regarded them as "advanced," but in the spring of that year the Peruvian Air Force purchased thirty, intending to use them as fighters. Fleetway bought a few as U.S. government surplus and flew them to Trenton, New Jersey, where they were rebuilt, fitted with bomb racks, and .50-caliber machine guns, and flown to Miami.

In Miami, Jack Ford gave his newest employee her first flight check in a military aircraft, and the next morning they began ferrying two T-6s south to Lima. From the Florida coast they flew two and a half hours to Camaguey, Cuba, the first refueling stop. From there it was a shorter leg to Kingston, Jamaica. At Kingston they were to remain overnight; they would make the last leg of the trip the next day.

That night at a little hotel in Kingston, Jack unrolled a navigation chart. He put a finger on the little island of Jamaica and drew it slowly down the paper until it reached the northern coast of South America. "Your distance from here to Barranquilla, Colombia, is five hundred and twenty miles over empty ocean—no navigation aids, no islands for checkpoints. The prevailing winds are against you, and Barranquilla is four hours away. Your fuel supply is exactly four hours and fifteen minutes." He looked up from the map, no trace of a smile. "Be right or be wet."

But it was Jack, not Jerrie, who ran into trouble. The next afternoon, when they put down in Cali, Colombia, his wheel caught in a chuckhole in the runway. He was fine, but the plane was too damaged to fly, and they both knew one of them would have to wait with it until the right parts could be shipped in from the States. Jerrie offered to stay behind while he delivered her plane, but Jack shook his head. "You brought it this far," he said. "You deliver it."

A few hours later she was airborne again, alone this time, with the verdant northernmost Andes rising on either side, their peaks shrouded in clouds. She crossed safely into Ecuador, where she touched down in Guayaquil to refuel. There she ran into trouble of her own.

She had just switched off her engine at the ramp when she was approached by a number of uniformed men, all carrying rifles and unholstered pistols. She showed them clearances from the U.S. State Department and the Ecuadorian embassy in Washington, but they ordered her to leave the plane and then, under armed guard, drove her from the airfield to what seemed to be an Army post. She was placed in a room that apparently served as a jail, the uniformed men tried to interrogate her, and gradually she began to piece things together. Ecuador and Peru were at war, and so when she landed in an airplane with Peruvian insignia, bomb racks, and .50-caliber machine guns, it was deemed prudent to detain her. She remained in the room for twelve days. At last the Ecuadorans determined that, as the purchase agreement would not be completed until the plane had been delivered, the T-6 technically still belonged to Fleetway, which made it the private property of a U.S. citizen. Jerrie was taken out of the room and back to the airport. The men with the pistols and rifles stood along the runway and waved as she took off.

Two refueling stops later, she landed at Las Palmas, headquarters of the Peruvian Air Force in Lima. Word of her difficulties in Ecuador had reached Peru, and she was greeted like a hero, escorted by military leaders to the Old Gran, the finest hotel in Lima. Jerrie was gratified, but she was also tired. She hardly had time to notice the luxurious appointments of the suite before she dropped onto the bed and fell asleep.

In 1952 some of Jackie Cochran's friends suggested that a comprehensive record of her achievements was in order, and the next year she published an autobiography called *The Stars at Noon*. Beneath her name on the title page was an addendum, reading, "with Floyd Odlum as Wingman." That same year, an Associated Press poll of newspaper editors named her "Woman of the Year in Business." They chose her for that title again in 1954.

1954–1959

Within six months, Jack Ford had revised his opinion of women pilots enough to put Jerrie Cobb in charge of all of Fleetway's South American operations. As she got to know Jack, Jerrie began to revise her opinion of him too. They became friends, and then they became more than friends. It was not a standard courtship. In early fall 1954 Fleetway began ferrying surplus B-17s to France as part of a lend-lease program, and Jerrie was assigned the North Atlantic route. She made six deliveries before she was ordered to Calcutta to retrieve a C-46 troop transport. As often as not, Jack was off in one direction and Jerrie in the other. Their courtship included trans-Atlantic phone calls and, less often, stolen moments between deliveries. Sometimes their flight paths would cross en route, and together they would walk the streets and take in the sights of London or Lima or Lisbon . . . or Wichita. They spent Christmas of 1954 in Oklahoma with Jerrie's family.

In the new year Jerrie returned to the South American run, on which she would pilot a Consolidated PBY, a plane that had been used for search and rescue by the Navy. It was a great twin-engine amphibian, with a fuselage that was half boat. It was government-surplus, and because the boost systems had been removed, it took enormous strength just to move the controls. Jerrie and a flight engineer flew it south to Paraguay. Her instructions were to set down at night in the Paraguay River at a point upstream from

Asunción. She did. Soon a launch came alongside. The engine idled and lines were thrown from the boat around the wing struts and the launch was pulled against the plane. A man said, "Capitán Cobb, you must take off at once for Buenos Aires."

Jerrie was tired, and there was no mention in her orders of any trip—at once or otherwise—to Buenos Aires. She politely suggested that he find one of his own pilots to make this trip. "But, Capitán." He shrugged. "We do not have any pilots who can fly this airplane." Jerrie responded, "Then I'm sorry, but the answer is still no." The Paraguayan official became almost frantic, and tried to explain: "There is a great man in Buenos Aires, a figure of much importance. His life is in danger."

Jerrie was so tired that a trip to Buenos Aires seemed likely to put her own life in danger, but the entreaties continued, so at last she told him to find one of his own military pilots, and she would teach him to fly the PBY. To this, he at last consented. The launch departed, Jerrie and the flight engineer waited, and a few hours later the launch returned bearing a military pilot. Introductions were made, Jerry showed the pilot the controls, and gave him the fastest check ride in history. A few minutes later the PBY took off, and Jerrie and her flight engineer were asleep in the backseat of the car that was taking them into Asunción.

The next morning it was all over the radio, and by then her Spanish was good enough that she could follow the whole story. A revolution in Argentina had forced the exiled Argentinean leader to find sanctuary on a Paraguayan gunboat in the Buenos Aires harbor, but the harbor was blockaded by Argentinean warships, and the new Argentinean government was suggesting that the exiled leader be returned to his native soil. The exiled Argentinean leader had mysteriously disappeared off the Paraguayan gunboat, possibly by airlift. The figure of much importance whom Jerrie Cobb had politely declined to rescue was none other than Juan Perón.

In 1955 Jackie Cochran decided to run for the House seat from her district, in California's Coachella Valley. She wanted a place in Congress, she had considerable experience overcoming the odds, and she had the support of the president himself. Eisenhower was a family friend, as was Senator Lyndon Johnson of Texas. For nine months, she campaigned hard, flying herself to speaking engagements all over the district. She had her talks written out on notecards, but most often she would abandon the notecards and just talk. After one such presentation, a man stood up and said he had a question. He was a teetotaler, he said, and what he wanted to know was, did Jackie drink? She said, "Yes, I drink. But two years ago I flew an airplane faster than the speed of sound, so I guess I don't drink too much." That was as diplomatic an answer as candidate Cochran was likely to give.

Floyd, who understood that politics involved diplomacy as well as vision, knew that the vicissitudes of democracy did not suit his wife's management style, and he suggested as much, as gently as he could. But by then the crowds had thinned, and Jackie herself was no longer so confident. On Election Day, when the returns came in, no one was surprised that she had lost.

By 1955 Jerrie had put in two good years at Fleetway, and Jack had asked her to marry him. But their relationship had reached an impasse. It seemed to Jerrie that asking Jack to settle down was like trying to cage an eagle. They continued to talk of marriage, but she was beginning to fear that it wasn't right, that mutual love of flying, even mutual love and respect for each other, was not enough. It was clear that she and Jack wanted to lead different kinds of lives. She took a two-week leave of absence, and by the end of it, she had made her decision. She was leaving Fleetway. She returned to the Midwest, and took a job as chief pilot for the Executive Aircraft Company in Kansas City.

Jerrie also returned to racing, and in the next few years she found opportunities to set records. On May 25, 1957, she flew an Oklahoma-built Aero Commander from Guatemala to Oklahoma City, covering fifteen hundred miles in eight hours and five minutes. The previous distance record in that weight class had been held by a Russian Air Force officer flying a Yak-II. For Jerrie, it was the smallest of victories on the outer edge of the battlefield of the cold war, but it was hers.

But it was the second flight that changed everything. On the second flight, Jerrie took the same Aero Commander to an altitude record. She climbed to 30,361 feet. It was as high as she had ever flown, and very nearly as high as that aircraft could fly. She climbed high enough that the sky turned dark blue. She saw no horizons, no boundaries. She could lean forward in the seat and look up through the Plexiglas canopy, stare at that bright blue emptiness and know that beyond it was space itself.

In 1959 Jack Ford was killed in the explosion of an airplane he was ferrying across the Pacific, just after take-off from Wake Island. A few weeks later, *Flight* magazine published a tribute which read, in part, "He was as capable as a goodwill ambassador as he was in delivering airplanes . . . He was one of the completely dedicated pilots . . . so intent in their love of flying and its perfectionism that they become out of place and impractical anywhere but in the air." It was all true enough, but to Jerrie Cobb even these words were inadequate.

It is curious that the subject of flying—maybe the most naturally poetic, most inspired and inspirational physical endeavor of the new century—is represented in many poetry collections only by a solitary work, by a pilot named John Magee. Magee flew a Spitfire for the Royal Canadian Air Force during World War II. He was killed in the fierce fighting of the Battle of Britain at the age of nineteen, and his contribution to belles-lettres is a short poem about flying an airplane. It is called "High Flight." Pilots on the whole are not poetic types, but after the war, Magee's poem started appearing in military airbases, scribbled or typed, stapled on bulletin boards next to flight rosters, or hung on the wall among faded photographs of old airplanes. And because military flyers, after they were discharged or retired, began to work for private firms, it wasn't long before the poem began to be displayed at civil airfields too. The experience it described wasn't something anyone talked about exactly, and now they didn't have to, because it was right there on the bulletin board.

Jerrie Cobb had seen this poem often enough, and she knew

exactly what Magee meant. She had felt the same exhilaration and she had held the same reverence. And when she was flying, when she was high in that sunlit silence, she felt it most of all . . .

Oh, I have slipped the surly bonds of earth
And danced the skies on laughter-silvered wings
Sunward I've climbed, and joined the tumbling mirth
Of sun-split clouds—and done a hundred things
You have not dreamed of—wheeled and soared and swung
High in the sunlit silence. Hov'ring there,
I've chased the shouting wind along, and flung
My eager craft through footless falls of air.
Up, up the long, delirious, burning blue
I've topped the windswept heights with easy grace
Where never lark, or even eagle flew
And, while with silent, lifting mind I've trod
The high untrespassed sanctity of space,
Put out my hand, and touched the face of God.

PART III

20

PROJECT ASTRONAUT

In the early 1950s America's future in space was only vaguely defined, but it was clear that there would be one. *Collier's* magazine had published a series of paintings by artist Chesley Bonestell which depicted a magnificent space station like a huge silver bicycle wheel spinning slowly among floating winged spaceships, the blue earth far, far below. In the popular mind this would be just the beginning. Later, there would be other space stations, and then voyages to the moon and to the planets.

Hollywood, of course, had gotten there first. In the fifties there were a lot of films about travel to other planets. This was a natural enough development, as the subject offered all the thrills of the western and all the novelty of the space age. Crew-cut young cowpokes riding herd on the stars. But the writers of these films had a problem. Nobody's going to see a movie without . . . *the girl.* The writers knew they had to have the girl for the movie poster, but they weren't exactly sure what else she might be good for. Gradually though, they found that they needed her for a lot of things. They needed her to hear explanations of important scientific facts ("A light-year, my dear, is the distance light travels in a year."), they needed her to listen wide-eyed to B-movie philosophy ("Perhaps, someday, man will venture among the stars . . ."), they needed her to make the mistake that gets them all in trouble ("Hmmm . . . I wonder what this cute little doohickey does . . ."), they needed her to run screaming from bug-eyed alien monsters ("Aaaaaeeeeeiiiihhhh!!!!"). And, of course, they needed her to get rescued ("Oh, Mike!—I thought I'd never see earth again . . . or *you* again"). So there'd always be the girl. The only real difficulty was finding an excuse to get her aboard the spaceship to begin with. Maybe she'd be the attractive daughter of the curmudgeonly professor who refuses to go anywhere without her ("Oh, Daddy—you and those old test tubes!"). Maybe she's a kind of tomboy who always wanted to go to Mars and so becomes a stowaway ("Well, what have we here?" "Never mind her, Mike. We've got bigger problems: Brace for a meteor storm!"). And maybe, just maybe, the girl actually has credentials. (The raised eyebrow: "A girl?" And then the reassurance: "She happens to be the world's foremost authority on extrasolar floombergeist.") Before the movie was over, the foremost expert would remove her horn-rimmed glasses to reveal large soulful eyes, and take that pencil out of her hair, which would then cascade beautifully over her shoulders.

As silly as this picture of womanhood was, at least there was a gesture at gender balance, a recognition that a female element was needed, even if that element was represented by an ornament. This picture of womanhood, in comparison to the reality of America's venture spaceward, was downright progressive. The reality was a boys' club. Or to put it more precisely, several boys' clubs.

America took its first tentative steps into the space age in 1948, when military and university scientists began a series of experiments designed to see whether living organisms could survive the radiation and weightlessness of space. Their leader was a brilliant and charismatic engineer named Werner von Braun. He was an engineering visionary, and his ideas were nothing if not ambitious. He imagined fleets of spacecraft, whole colleges for space exploration, and eventually, a civilization among the stars. In 1956 he was named chief of the Guided Missile Development Division at the Army's Redstone Arsenal in Huntsville, Alabama, where he began to develop Project Orbiter, a plan to launch an earth satellite.

Von Braun had worked for Hitler during the war, and to the average American, there was something slightly wrong with von Braun's prominent position in the nation's space effort.* It did not help that the Redstone missile being tested for Project Orbiter looked a lot like the German V-2, of which it was in fact a direct descendent. The Eisenhower administration, appreciative as it was of German engineering genius, did not want the first American satellite to be launched by the rocket that had terrorized Britain. Further, it wanted to maintain a separation between satellite research and military operations. And so in 1955 the National Security Council ended Project Orbiter. Although von Braun could continue to develop tactical battlefield missiles, the only U.S. satellite project would belong to the Naval Research Laboratory, which had been developing a small scientific satellite called "Vanguard" as the United States' contribution to the general sharing of scientific findings called the 1957–58 International Geophysical Year.

While most Americans in the late 1950s believed that space travel was inevitable, they still regarded it as part of a vaguely defined future. Then on October 4, 1957, came the news. The Russians had launched an earth satellite, an *artificial moon*. They had a name for it: *Iskustvennyi Sputnik Zemli* or "Fellow Traveler of the Earth." Anyone could stand out in the backyard after supper and look up into the cool October twilight and actually see it. It looked like a fourth-magnitude star, about as bright as one of the stars in the Big Dipper. But it was slowly moving. The satellite was a blow to American complacency and an astonishing display of technical prowess. It was also new cause for fear.

In 1957, thoughts of a nuclear war rippled through the fabric of American life. It had been six years since the Federal Civil Defense Administration had begun to establish community shelters and instruct the public on how to protect itself in the event of a nuclear attack. In schoolrooms all over the country the teacher would lay

*In a few years there would appear a popular biography of von Braun called *Reaching for the Stars*. And a number of editorial writers would add a subtitle: ". . . sometimes I hit London."

down her pencil and issue instructions in a quiet, almost unnaturally calm, voice. The students would push back their chairs, get down on their hands and knees, crawl under their desks, scrunch up with their heads between their knees, and shut their eyes tight. What was particularly disturbing was the routine of it, the way it was becoming a part of the rhythm of an ordinary American childhood—math and English and basketball and piano lessons and preparation for nuclear war.

Films and pamphlets from the Civil Defense Administration had suggested that an attack would come from manned bombers. And so in October 1957 the average American, conditioned to associate bombs with aircraft, did not immediately grasp the military implications of that tiny point of light. Khrushchev made the connection clear. He claimed that the missile that had launched Sputnik could carry a hydrogen warhead halfway around the world in forty minutes.

The Navy's Vanguard was scheduled to launch on December 6, but its missile was new and untested. Von Braun's Jupiter missile, on the other hand, was of a proven design, and had been used in various configurations since the forties. The White House and Pentagon realized that von Braun offered the best chance to answer the challenge from the Soviets. The Pentagon gave him the green light. But he would need eighty days. In the meantime, the Vanguard was still scheduled to be launched on December 6. And so Americans on that Friday morning watched the image of a pencil-thin rocket on the screens of their twenty-one-inch RCA Victor black-and-white televisions. There was a dramatic countdown to zero, and then a roar of flames and smoke from beneath. The pencil-thin rocket lifted slowly off the pad a few feet, balanced for a moment like a broomstick on a finger, and then began to sink and fall sideways. Before it fell farther, almost mercifully, it exploded into flames. The third stage and its payload, a three-pound aluminum sphere about the size of a grapefruit, were tossed away from the explosion and fell to the ground. The sphere, lying there in the marsh grass, was still transmitting.

But on January 31, 1958, von Braun and his team launched his Jupiter missile. The Explorer satellite it placed in orbit weighed only thirty-one pounds, but it worked. America had its own artificial moon. If the rocket that launched it was a battlefield missile of German design, who cared? Suddenly questions of technological ancestry seemed beside the point. Enough time had been wasted looking backward.

In the early spring of 1958 the Soviets were still far ahead. It was obvious they were preparing to launch a man, and it was clear that unprecedented measures were necessary if America were to launch a man first. The Senate Preparedness Committee recommended the establishment of an independent space agency. The American Rocketry Society called for a civilian space research and development agency. The National Academy of Sciences proposed a National Space Establishment. By April, members of Congress had introduced twenty-nine bills and resolutions relating to the reorganization of America's space efforts.

The purpose of manned spaceflight—the why of it all—would be reflected in who was running the program. The National Advisory Committee for Aeronautics (NACA), probably the world's foremost aeronautical research organization, considered the answers to be scientific interest, national prestige, manifest destiny and national defense, more or less in that order. Its own proposal recommended that any national space program should involve itself, the National Academy of Sciences, the National Science Foundation, and the Defense Department.

To many in the Air Force, questions about "proper administration" and "why?" were late-night talk for philosophers and social theorists. The Air Force regarded space travel as the natural and obvious successor to air travel, and outer space as its own eminent domain. One lasting legacy of Hap Arnold was a continuing program of research and development, by the 1950s institutionalized as the Air Research and Development Command (ARDC).* In 1956 the command was headed by General Thomas S. Power. When the Air Force was denied the funds to orbit an unmanned satellite in 1955, Power's staff began to work quietly on feasibility

*In 1945 Arnold had asked physicist Theodore von Karman to head a scientific advisory group to chart the future of Air Force research and development. Missiles were nothing new to Arnold. As early as 1919 he and Orville Wright had observed the successful launch of the Air Service's first guided missile—a "flying torpedo," they called it. By 1944 the German V-2 had convinced Arnold that missiles represented the future of air weaponry.

studies for piloted spacecraft. A few days after the launch of Sputnik, an Air Force–appointed committee of academic and corporate scientists and Air Force officers had recommended that the United States adopt a unified space program under Air Force leadership. In fact, Power simply assumed that ARDC's several space projects would be funded. The blue suiters lacked neither ideas nor ambition. In 1958, when no one had been higher than a few miles into the upper atmosphere, they were already proposing an eleven-step "Manned Space Flight to the Moon and Return." They were serious. First, an automated capsule would be launched into earth orbit. It would have an ape inside. Then, another automated capsule would be orbited, this with a man inside. And if all went as planned, a new, larger capsule would be sent around the moon with an ape inside, followed by a second capsule holding a man. Finally, a man—presumably an Air Force officer—would land on the moon. This would occur in 1965.

In 1958 the ARDC began soliciting proposals from aircraft manufacturers for manned spacecraft systems. In January they held a closed conference at Wright-Patterson Air Force Base, during which eleven aircraft firms presented proposals for manned space flight. Most of the proposals were like that from McDonnell Aircraft Corporation in St. Louis. It suggested a method called "ballistic flight," meaning the spacecraft would not so much fly or glide, as be shot into space on a missile, and simply fall back to Earth. In no sense would the person inside control the spacecraft. In fact, in everything but name he would be a passenger, a "recoverable biological payload" not unlike the mice and monkeys von Braun had been launching out in the New Mexico desert. All the spacecraft systems would be automatic because nobody knew for sure whether a man could stay conscious, let alone operate controls when he was being subjected to high g-forces and made weightless.

Not all these proposals, though, were for ballistic flight. Northrop Corporation had a proposal based on a preexisting joint study by the Air Force and NACA, a proposal that seemed a little more dignified, a project more worthy of the proud Air Force tradition of men who controlled their craft. It was a strange winged machine that would be launched on the still-unbuilt Titan rocket. It would be launched into space on a missile, but its pilot would return by skipping off the atmosphere the way a flat stone skips off the surface of a pond. Because this particular trajectory, a series of skips, was called "dynamic soaring," the Air Force called this machine the "Dyna-Soar." The pilot would slow the Dyna-Soar with each successive skip; by the last one it would be subsonic. And the pilot would land it like an ordinary glider.

At the conference at Wright-Patterson there were other proposals for piloted flight. One was from North American Aviation, the builder of the X-15. The X-15 was a rocket plane, a descendent of the X-1, the ferocious little machine in which Chuck Yeager, in 1947, had become the first human being to fly faster than sound. In the North American plant in Los Angeles on October 15, 1958, there was a roll-out ceremony. The hangar door opened and spectators saw a little service tractor pull out an evil-looking missile about fifty feet long. It began with a nose cone. Above and behind it were two slit windows made of glass three inches thick, and from there on back it was a long fuel tank with two stubby wings, a tail section, and finally, an engine exhaust nozzle. Its glossy black skin was a nickel alloy that could withstand temperatures in excess of twelve hundred degrees Fahrenheit. It was a pure research craft, designed for aerodynamic studies at six times the speed of sound and altitudes greater than 200,000 feet. It was not designed to carry weapons, and it never would. But when it was rolled out of that hangar, it looked like something out of a Russian's worst cold war nightmare. Like the X-1, the X-15 would not take off from the ground. Instead, it would be carried aloft beneath the wing of a B-52. At 10,000 feet it would simply be dropped, and its pilot would light its engine and launch it up through the stratosphere. When its fuel tanks were exhausted it would keep going upward on sheer momentum, like a great hurled javelin. Its long arcing trajectory would take it above the blue rim of anything that might be called air, to the very edge of space. The highest point would be over fifty miles, so high that ailerons and rudders wouldn't work, so high the pilot would have to control it with small hydrogen-peroxide thrusters in the nose and the wings. At that point he would actually see the curve of the earth's horizon. And for a few glorious heartbreaking moments before the long glide back, the X-15 would have actually become . . . *a spaceship.*

In January 1958 the roll-out ceremony for the X-15 was eight months distant; the first actual flight tests out at Edwards would not begin until summer of the following year. Still, North American engineers were already proposing a plan whereby the X-15B—a lighter version—would be launched on a three-stage rocket, orbit the earth once, and then return. And because slowing down was so difficult, the pilot wouldn't even bother. The rocket-plane would come thundering through the stratosphere and at a predetermined altitude the pilot would eject from the cockpit, then separate himself from his tumbling ejection seat and pull the ripcord that would open his parachute.* And long before his feet touched earth, his rocket plane, still traveling at an incredible speed, would crash into an unpopulated area within an Air Force testing range somewhere in the desert. Boom. The only thing left of the million-dollar craft would be a smoking crater. But that wouldn't matter. Because when the reverberations of the explosion had died away and the smoke had cleared, someone would be standing there on the desert floor. And it wouldn't be just any someone. It would be an American who had flown into outer space.

In fact, if Power had his way, it would be an Air Force officer. But the X-15B could not be ready before the summer of 1960, and the Dyna-Soar would not be ready before 1963. The other projects presented at Wright-Patterson would not be completed in less than two years. As far as Power was concerned they might as well have said two hundred. Suddenly the imperative was to get an American, just one, into space—*real* space—as soon as possible.

Thus although the Air Force did not abandon its long-range plans, it was also thinking of the short term. On February 27, General Curtis E. LeMay, Air Force vice chief of staff, ordered the Air Research and Development Command to devise a way to put a man into Earth orbit quickly. So Power and his staff drafted a proposal for a project they called "MISS," for "Man In Space Soonest," which about said it all. This would be no glistening winged spaceship from von Braun's future. It would not be a hypersonic glider, or even a kamikaze X-15. It would be the reentry

*Ejection at extreme speeds and altitudes made possible a wide variety of violent deaths. One type of accident, termed "premature ejection," elicited some dark jokes, but there was little funny about it. The pilot was supposed to curl up into a ball and jettison the canopy. If he was not curled into a ball, the ejection could amputate both legs above the knees, and if the canopy were not jettisoned first he would be shot against it, snapping his spinal cord, maybe breaking his back.

vehicle of a nuclear missile with a man inside, where the warhead was supposed to be. It was a brute metallic cone about nine feet high and five feet across its heat shield base, a design chosen because upon reentry it was intended to come in backward, heat shield first, and actually slow down. It would be rocketed into space by a ballistic missile. And it would not fly back to Earth. It would simply fall.

Meanwhile, Eisenhower had asked the Department of Defense to create an organization to review and manage space projects proposed by the Air Force and others. It was called the Advanced Research Projects Agency (ARDC), and Roy W. Johnson was named its director. By early summer, Roy Johnson and his staff were evaluating a number of proposals for manned spaceflight.

The Navy Bureau of Aeronautics, still smarting from the memory of its little Vanguard radio transmitter beeping in the marsh grass, now proposed an orbital mission in a cylinder with parts that would inflate to become a delta-wing glider with a rigid nose section. It was called "MER," for "Manned Earth Reconnaissance." Most said it was far beyond the abilities of current technology. The Army, flush with the success of von Braun's Explorer satellite, had a proposal called "Army Adam." The idea was to put a man inside a sealed capsule, put the capsule inside the nose cone of a Redstone ballistic missile, and launch it to an altitude of about 150 miles out over the Atlantic. Near the top of its parabolic trajectory it would separate from the Redstone, fall a few miles, then use a parachute to land gently in the water.*

There were others in the race, too. Frederick de Hoffman was the founder of a San Diego firm called General Atomics, a developer of commercial nuclear reactors. In 1958 de Hoffman began working with Theodore Taylor, a veteran of the Los Alamos weapons programs, and Freeman Dyson, a theoretical physicist with the Institute for Advanced Study in Princeton. Their proposal, called

*In one way Army Adam was one of the most bizarre proposals ever to come over the transom. Its authors claimed it would prepare the way for "the transportation of troops by ballistic missiles." It wasn't clear how serious the Army was about putting a soldier inside the nose cone of an intercontinental ballistic missile and lobbing him over the Iron Curtain. In fact, the proposal's appeal to military objectives may have been insincere; von Braun himself was among the authors, and there is evidence that he believed the Army would only fund work with proven or promised military applications.

Project Orion, involved a spacecraft that would use hydrogen bombs to fuel something they called a nuclear-pulse engine. At timed intervals, a bomb would be ejected from the rear of the spacecraft, where it would detonate against a "pusher plate," a combination shield and bumper positioned between the explosion and the spacecraft. In July 1958, the Advanced Research Projects Agency agreed to sponsor Project Orion at an initial funding level of $1 million per year.

Relatively speaking, the Navy MER, Army Adam and Project Orion were late starters. The odds-on favorite was the Air Force. If the Air Force slowed down enough to contemplate the competition, it would have regarded MER and Orion as unrealistic, and the Army Adam as typical grunt thinking—spacecraft not as the successors of aircraft but of ordnance. (How to put a man in space the Army way? Drop him inside a cannon barrel and light the fuse.) But the Air Force did not slow down. There was work to do. In the summer of 1958 the ARDC revised the Project MISS proposal and Roy Johnson rejected it. The ARDC revised it and submitted it, and Roy Johnson rejected it again. They had a new development plan every few weeks. On July 24, they submitted the sixth, which claimed that an Air Force officer could be in orbit before June 1960 at a cost of $106 million. Already they had begun studies for a worldwide tracking network, heat shields, and alternate launch vehicles. Everything was gearing up for the big push. Everything was ready. Well, almost everything. The capsule designs were uncertain, but invitations to contractors could be mailed within twenty-four hours. . . .

To the lay public these proposals and debates involved rather arcane matters, and if the newspapers reported them at all, it was on page twelve or nineteen in a three-paragraph column next to the furniture ads. Maybe because of those pictures of space stations in *Collier's* magazine a few years before, maybe because of half-remembered stories of H. G. Wells and Jules Verne, many Americans were disappointed to realize that space might serve only a military purpose after all, that we could not see beyond our need to protect ourselves, that we could not dream larger dreams of what we might accomplish in the largest of frontiers. The spaceships in the movies and on the covers of science fiction paperbacks had graceful curving lines. They looked like beautiful arrowheads or futuristic sculptures or even Christmas ornaments. But the spaceships that were beginning to appear in the newspaper photographs in the spring of 1958 didn't really look like spaceships. They looked exactly like ballistic missiles.

So despite the urgency of the situation, and despite the fact that the military had a great many rockets on the launchpad or in development, there was growing sentiment in Congress to ensure that space activities had nonmilitary ends. The leading proponent of a peaceful path to the stars was Eisenhower himself: He had envisioned Vanguard as a kind of scientific instrument. On July 29, 1958, he signed into law the National Aeronautics and Space Act. The relevant passage began "The present National Advisory Committee for Aeronautics (NACA) with its large and competent staff and well-equipped laboratories will provide the nucleus for NASA. The NACA has an established record of research performance and of cooperation with the armed services. The coordination of space exploration responsibilities with NACA's traditional aeronautical research functions is a natural evolution . . . [one which] should have an even greater impact on our future."

For months now the ARDC had seen it coming; Eisenhower had sent the bill to Congress in early April. Still, it was hard for the blue suiters to believe the high ground would be given to researchers. But it was. With a stroke of the pen, eight thousand NACA scientists and engineers were made employees of a new organization called the National Aeronautics and Space Administration (NASA). NACA facilities, policies, and advisory committees were transferred to NASA, and NACA laboratories became NASA Research Centers. Much else that involved America's work in space—the Navy's Vanguard, the Army's Explorer, the services of the Jet Propulsion Laboratory, several rocket engine programs including an Air Force study contract with North American for a million-pound thrust engine, the findings from the International Geophysical Year, five space probes, three satellites and $117 million in appropriations from the Defense Department—were given to NASA. The Air Force was sitting at the great banquet table of government contracts staring at an empty plate. In late August the plate itself was removed. Eisenhower assigned NASA the specific responsibility of carrying out a manned space flight. The Department of Defense, which had set aside $53 million for the Air

Force's MISS and whatever idea it had that week, transferred every cent to the new space agency.* As far as the Eisenhower administration was concerned, the best answer to the *why* of manned space flight was scientific interest, or, in the words of the National Aeronautics and Space Act of 1958, "the expansion of human knowledge of phenomena in the atmosphere and space." And so the new space agency, which would be formally established in October of that year, was civilian.

In November 1958, twenty-seven former NACA engineers and eight administrators organized themselves into a NASA committee whose specific goal was nothing less than the achievement of manned spaceflight at the earliest possible date. They called themselves the "Space Task Group." On the matter of manned space flight, the question of *why* had been answered, and the question of *how* was being addressed. In the fall of 1958, people answering the *how* began to think about the question of *who*.

By early November the aeromedical team, the Committee on Life Sciences, and NASA headquarters had determined that the passenger of the space capsule would be between twenty and forty years of age, weigh no more than 180 pounds, and have a degree in science or engineering or the equivalent. The sponsoring program was to be called "Project Astronaut."† The job would be a civil service position, and it would be advertised as such; the plan was for open applications. Test pilots were the preferred candidates, but anyone who met the short list of criteria would be considered. The task group presented its recommendations for astronaut selection to NASA administrator T. Keith Glennan, and

*On September 11, the Air Force submitted a seventh version of its MISS proposal to the Advanced Research Projects Agency, which did not approve it. The Air Force's Dyna-Soar project would be canceled in 1964. Some joked that the Air Force, with its missed MISSES and extinct Dyna-Soars, was in desperate need of help from Madison Avenue. In December 1963 Secretary of Defense McNamara announced that the Air Force would be assigned a "Manned Orbiting Laboratory" (MOL), in which Air Force astronauts would work for thirty-day periods. That project was canceled too; in June 1969 Air Force pilots in training for the MOL missions were transferred to NASA.

†The name was a variation on "aeronauts," a term that had been used to describe early balloonists, and "Argonauts," the legendary Greeks who sought the Golden Fleece. Literally, the word means "star voyagers."

Glennan passed them along to the president. Eisenhower was no stranger to personnel issues, and he knew an open application process would invite inquiries from every loose screw in the country. He made it clear that he wanted military test pilots for the simple reason that they had already undergone rigorous screening, and their records were on file in Washington.

During the first week in January there was another series of meetings at the task group headquarters at Langley Research Center in Virginia. In attendance were Robert Gilruth, NASA's deputy administrator, Warren J. North, a former NACA test pilot, and the task group's project manager, George Low, a soft-spoken engineer now acting as a liaison between NASA's field activities and its headquarters. Although a few expressed regrets that volunteers for a civilian organization would be drawn only from the military, they also understood Eisenhower's reasoning, and were grateful that at least some of their work would be made easier.

Gilruth, North, Low, and the others did not know what might happen up there. No one did. No one really knew what kinds of human bodies, or human minds, were best equipped to undertake such a mission. But now they began to make their best guesses. They had regarded the earlier civil service description as overly vague, and in a matter of days reduced it to a straightforward, seven-item list. The first three items remained from the earlier description: The astronaut would be less than forty years of age and in excellent physical condition, because the physical stresses of spaceflight were likely to be great; and he would be under five foot eleven inches because the capsule could accommodate no one larger.* But the remainder of the list was the task group's answer to the rather theoretical question simmering beneath the surface of the manned spaceflight studies of the past year. Would the astronaut be merely a recoverable biological payload, or would he be the single most important spacecraft system? Would he be passenger or pilot? The task group's answer, in emphatic agreement with Eisenhower, was that he would be a pilot. Moreover, he would be a very good pilot. He would have a bachelor's degree in science,

*At the place where it joined its nose cone, the Redstone had a diameter of only 74 inches. It was determined that this diameter could accommodate a sitting person, accounting for a pressure suit and capsule bulkhead, only if he or she were no taller standing than five feet, eleven inches.

engineering or its equivalent, as a guarantee of his ability to understand complex systems. He would be a graduate of a test pilot school because the associated degree implied a knowledge of electrical, mechanical and hydraulic systems and training in evaluating new designs. He would have fifteen hundred hours—in most cases this meant ten years—because this was considered a reasonable lower limit of necessary flight experience. Finally, he would be qualified in jets because split-second actions and decisions might be required to save a spacecraft, a mission, his own life, or all three. It went without saying that the task group regarded spacecraft as the next generation of high-performance jet aircraft.

During those meetings at Langley in the first week of the new year, one of the men in attendance was Brigadier General Donald D. Flickinger. At fifty-one, Flickinger was widely regarded as one of the few experts in the young field of aeromedical research. His studies in crash survival and flying fatigue had led to the establishment of the position of flight surgeon. Recently, he had served on a NACA working group to study the physiological problems of manned spaceflight. He had been the director of Human Factors in the Air Force's Research and Development Command during the period it was evaluating MISS. In 1958 he had informed his superiors that from a medical standpoint, a man could probably survive a trip into space.

Nonetheless, Flickinger knew that the astronaut, no matter how accomplished his experience as a pilot, would have to be extraordinarily fit. Certain physical defects would be manifest only under stresses peculiar to high-altitude flight. The heart of an otherwise healthy man, for example, might have tiny congenital openings between its right and left sides. Under normal circumstances he might live a long, healthy and unremarkable life, but if the same man were sitting in a pressurized cockpit at a great altitude, and that cockpit suddenly decompressed, those tiny congenital openings could kill him. It could happen in an ordinary commercial airline. And every expectation was that spaceflight would be more demanding: an astronaut might undergo sudden decompression too, but he would certainly undergo tremendous g-loads and vibrations, and an abrupt shift to weightlessness. Then the sequence would occur in reverse: tremendous g-loads and vibration, heating to maybe 130 degrees Fahrenheit, and a violent impact on the water at thirty feet per second. Medical test subjects had experienced each of these conditions separately for brief periods, but it was impossible to experience them in combination or for any appreciable duration without actually going into space. As if all this weren't worrisome enough, there were unknowns. In 1959 there was a popular science fiction movie, *First Man into Space*, in which a handsome jet pilot, the "first man" of the title, accidentally flies through comet dust and returns to earth as 180 pounds of shapeless, bubbling ooze. It was only a movie, still . . .

Obviously, there would have to be a medical testing program. When Gilruth raised the matter of suitable facilities, two suggestions were the Army's Walter Reed Medical Center, and the Bethesda Naval Hospital. But in subsequent meetings it seemed that a civilian facility might be more appropriate for a civilian program. Besides, launching a man into outer space in the nose cone of a missile was a fairly harebrained scheme, and there were fears that no military test pilot would risk sidetracking his career on it. A civilian facility would be under no obligation to release test subject records to the services, and so if the doctors discovered an anomaly, no military career need be forfeited. Flickinger shrugged and said, "Send them to Lovelace. They're used to keeping secrets."

The Lovelace Foundation for Medical Education and Research in Albuquerque had been established in 1947 by W. Randolph Lovelace II. It was supported by the older Lovelace Clinic, founded twenty-five years before by Lovelace's two uncles. By 1959 it was a highly respected institution. And Flickinger was right. They were used to keeping secrets. Although the clinic was a civilian facility, in 1958 its Department of Aerospace Medicine was headed by retired Air Force General A. H. Schwichtenberg, and it had already done various studies for the Air Force, and performed medical tests on civilian U-2 pilots then flying secret high-altitude reconnaissance missions over the Soviet Union.

As it happened, the task group was regularly reporting to the nine medical specialists who made up NASA's Special Advisory Committee on Life Sciences. On November 21 of that year, Glennan had appointed W. Randolph Lovelace as chairman of that committee "to advise NASA on the role of the astronaut and all related considerations." In 1958 Randy Lovelace was a tall man with dark eyes, a quiet manner, and credentials as impeccable as they were unique. During the war he had been chief of the Aero-Medical Laboratory at Wright-Patterson. Later he became a brigadier general in the Air Force Reserve, and had been awarded a steady flow of decorations, including the Distinguished Flying Cross, the Air Medal, the Legion of Merit, and three combat stars.* Randy Lovelace, at the behest of Flickinger and the Space Task Group, contacted General Schwichtenberg at the clinic, and the general devised a medical examination consisting of eighty-seven tests.

Meanwhile, in January 1959 the Space Task Group screened 508 service records and found 110 that met the minimum standards specified in the seven-item list. An evaluation committee arbitrarily divided them into three groups of roughly equal sizes, and invited the first to come to Washington for briefings and interviews. The task group was expecting a great many to decline. To the surprise of Low and others, twenty-four of the first thirty-five were willing to go further. The numbers for the second group were even better. As only six astronauts were needed, it was decided not to bother asking the third. In the middle of February, sixty-nine men reported to Washington for an initial series of written tests, medical evaluations, and psychiatric interviews. The number was reduced to thirty-six. They were invited to undergo physical tests at Lovelace Clinic, and mental and physical environmental tests at Wright Air Development Center in Dayton, Ohio. Thirty-two accepted and became candidates. The candidates were divided into groups of six, and at the Lovelace Clinic each group was examined and its members rated. There were examinations by an ophthalmologist, an otolaryngologist, a cardiologist, and a neurologist. There was also an examination by a surgeon, Randy Lovelace himself. Each of the thirty-two test-pilot specimens were prodded and poked and otherwise scrutinized for a week. Then each was sent to the Aeromedical Laboratory of the Wright Air Development Center for another testing phase, whose general direction was determined by

*Within aviation circles, Lovelace was best known for a particular feat of aeromedical research. In 1943 he had designed equipment to assist jumps at high altitudes, and he tested that equipment himself by jumping from the bomb bay of a B-17 at 40,000 feet—the second-highest bailout on record. The equipment was further developed, and became standard. For the history of aerospace medicine, that seven-mile fall was a significant moment. For Lovelace himself, it was important for another reason. It had been his first parachute jump.

Flickinger and the Space Task Group. It was called "stress testing," and included environmental studies, physical endurance tests, anthropometric measurements, and psychiatric studies. By late March the physicians, psychiatrists, psychologists, and physiologists at Albuquerque and Columbus had finished their work. They had tried to narrow the field to the required six, but they could not in good conscience get the number below eighteen. So back at Langley some of the task group—Donlan, North, White, and Gilruth—made the final selection based on the men's experience. Even then they were unable to reach the targeted number, and at Gilruth's suggestion, they recommended seven. On April 2, 1959, NASA announced the names of the men who had been chosen for Project Mercury. Exactly one week later, at a news conference at NASA's temporary headquarters in the Dolley Madison House in Washington, they were presented to the public.

On January 12, McDonnell Aircraft Corporation in St. Louis was informed that it had been awarded the contract to build the capsule, and by the end of March the McDonnell engineers were evaluating designs and finding subcontractors. An Atlas had flown its designed range of six thousand miles, and another had actually achieved orbit. By late April the men called astronauts were in training.

The contradiction between NASA's civilian status and the military background of its newest employees did not bother most Americans in the least. To most Americans it made perfect sense that the first astronauts would be military pilots. Who else would want to ride on top of a ballistic missile? Besides, the first American into space would be a hero of the first order, and military pilots were already heroes of the first order, a fact made evident in the attention given the graduation, in June 1959, of the first class of cadets from the U.S. Air Force Academy. This first graduation was pretty heady stuff—wave upon wave of future Jimmy Doolittles and Chuck Yeagers, the best and brightest of the proud young defenders of the American way. It was almost inevitable that *Life* magazine would run a cover story.

But *Life* wasn't interested in just the pilots; it was also captivated by their girlfriends. Subtitling its article "Girls and Weddings

Grace a Graduation," *Life* informed its readers that "The affair was . . . dressed up by the presence of the cadets' pretty girl friends, who came from all over the U. S. to bask in the romantic atmosphere of dress uniforms and military flourishes. Seventy-two of the new officers had announced their engagements and before the week was over fifty were married." One point of all this was to demonstrate that the future heroes of the skies hadn't let all this guts and glory business go to their heads so much that they'd forgotten their equally vital roles as husbands and fathers. But the other point, the more obvious one, was that the heroes of the skies were irresistible to women. The military pilot was the apotheosis of the masculine ideal.

But in the same issue of *Life* was another story about another kind of flyer, one who, in the grand old barnstorming tradition, made a living walking around on the wings of airborne airplanes. There was a picture of this flyer standing upside down, feet planted on the wings of a small prop plane, head pointed toward the earth thousands of feet below. The one-paragraph article that accompanied the photos identified her as Linda Earnhardt, a thirteen-year-old girl from a small town in North Carolina, and a modest challenge to the cultural assumption that women were always content to let men do the flying.

A similar cultural assumption informed the reaction to NASA's presentation of the seven Mercury astronauts. Their pictures were everywhere, and although in 1960 few people could tell which was Glenn and which was Carpenter, it soon became almost second nature to think of a handsome young man with a crew cut, *this* is what an astronaut looks like. Because most Americans had never entertained the possibility that there might be other types of astronauts, the image of the test pilot as astronaut seemed natural and inevitable. Even the members of the Space Task Group, for whom the association of the astronaut with test pilot had not been entirely automatic, had by 1960 grown accustomed to thinking of the astronaut as a test pilot. In his capacity as one of NASA's medical advisers, however, Randy Lovelace retained a slightly different perspective. As per NASA's request, he had focused his expertise on the physiological requirements of spaceflight. While he agreed that test pilots had been a good choice, he wondered if they were the best choice. In fact, it occurred to him that, from a medical perspective, the ideal astronaut candidate might be female.

Within the Space Task Group itself, few had the time to give much thought to such possibilities. The development of that seven-item list had been difficult enough, and as far as Low and North were concerned, it was sensible and defensible. Because spacecraft were the next generation of experimental, high-performance jet aircraft, the astronaut had to be a jet test pilot. Because there was no time for a lengthy screening process, that jet test pilot had to be military. It went without saying that all military jet test pilots were men. But as planning for Project Mercury began, some of the items on that list were questioned. Although no one was seriously proposing that the astronaut be a woman, some were still thinking he need not necessarily be a pilot. It had been particularly difficult for the missile builders—von Braun's team—to regard the spacecraft as anything other than the missile's payload section, and its occupant as anything other than a recoverable biological payload. Constructing a missile that was controlled not by technicians on the ground but by the payload section ran counter to a twenty-year tradition of design. Their arguments had the support of some of the psychologists. It would be unthinkably irresponsible to expect a man in outer space to act as a pilot, when no one knew if he would be able to function at all.

As the spacecraft was designed and constructed, there was a sort of ideological tug-of-war over the role of the astronaut. On one side was the Space Task Group's conception of the astronaut as pilot. On the other was von Braun's conception of the astronaut as payload. And caught somewhere in the middle were the capsule designers at McDonnell. At the Air Force conference at Wright-Patterson a year before, McDonnell engineers had proposed a ballistic flight, and a capsule that would be controlled automatically and/or remotely. In the spring of 1959 that conception survived in various recommendations and preliminary reports: "Primary control is automatic. For vehicle operation, man has been added to the system as a redundant component," and so on. But through the spring and over the summer, as the capsule design was being refined, other voices were heard. John F. Yardley, Max Faget's counterpart at McDonnell and chief developer of the Mercury capsule; a McDonnell human factors engineer named Edward Jones; and the task group's psychologist, Navy Lieutenant Robert B. Voas were all coming to believe that the astronaut could and should play a more active role. In August 1959 Jones and a colleague wrote a memo to Yardley recommending that if the capsule's automatic systems failed, the astronaut should be trusted to take over. By this time Jones was preaching to the choir. Yardley had already come to regard spacecraft as the successors not of missiles but of aircraft, and as far as he was concerned it wasn't just theory anymore, one could see it in the designs themselves. Soon enough, the astronauts themselves were adding their own much-considered opinions to the design process, and to no one's surprise they took the position that the astronaut was not a payload but a pilot. Glenn and Shepard and the rest could not imagine that space travel would be incapacitating. In fact, they expected that it would be much like flying high-performance jet aircraft. They had experienced g-forces eight, ten, and twelve times normal gravity and survived. They had even been weightless for a few seconds at the top of a parabolic trajectory. As for disorientation, they had recovered from flat spins where the world outside was revolving once a second. The suggestion that flying higher or being weightless longer would cause panic or apoplexy seemed mildly ridiculous. The astronauts lobbied for a rearrangement of the control panel, an escape hatch that could be opened from within, and a window.

In the matter of the role of the astronaut, the hottest issue was control. Control was the defining feature of the pilot. Remove his control, he's Spam. Return his control, he's a pilot again. To the general relief of von Braun's team, no one seriously proposed giving the astronaut the ability to throttle or otherwise direct the missile beneath him. So the only control that remained to be delegated was in the capsule itself. There were small attitude thrusters arranged around its surface which shot spurts of hydrogen peroxide so it could be turned, and braking rockets strapped to the heat shield which, when fired, would slow it enough that it could begin to fall earthward. Gradually, the McDonnell engineers gave in. Sort of. The design which had evolved by October 1959 had a rearranged control panel, an escape hatch that could be opened from within, and a window. And those thrusters and braking rockets, which would normally be controlled remotely from the ground, would include a way for the astronaut, in an emergency, to override them completely. There were also two partly automatic, partly manual types of control systems—one called "fly-by-wire," the other called "rate command." To hear all the talk about the new designs in the fall of 1959, anyone might think that Voas and Jones and the astronauts had won a great victory in the battle for the astronaut as pilot. In fact, they had won only a few minor embellishments of the original design. The primary control system of the Mercury spacecraft would still be automatic, and the astronaut would still be a redundant component. Even the hard-won control he would have if the automatic system failed wasn't much. True, he could fire one of the five-pound thrusters to turn the capsule this way or that, rotating it in the standard three directions of movement for aircraft—yaw, roll, and pitch. True, in an emergency he could ignite the braking rockets. But for most purposes he was still the inert payload of a ballistic missile, and he couldn't change his actual speed or trajectory anymore than a thrown football could change its speed or trajectory. Anyone still suffering under the delusion that the Mercury astronaut would be piloting his craft need only look at NASA's plans for the coming months. A chimpanzee would make the first flight. Its capsule, identical to the one that would carry the first human astronaut, would be controlled by onboard automatic systems and radio signals from the ground.

The pilot/Spam issue was not the only problem facing the McDonnell engineers in the fall of 1959. They were building what is euphemistically called an engineering marvel. It would have to keep a man alive against fierce radiation and harsh vacuum, and in temperatures that ranged from two hundred degrees above freezing to two hundred degrees below. It would have to withstand tremendous forces of acceleration, heat loads, and aerodynamic stresses. And because it would be launched into outer space by a Redstone and into earth orbit by an Atlas, it would have to weigh less than twenty-seven hundred pounds.

When most people saw the capsule for the first time they said, "It's so *small*." It looked a bit like a shuttlecock. It was a truncated cone of two layers of welded heat-resistant titanium, its base a convex ablative heat shield. It was just nine and one-half feet high, and barely seventy-four inches across at its widest point. Climbing inside, into the little space between the contour couch and the instrument panel, was like sliding under a bed. It was small because it had to be mated with a missile seventy-four inches wide, and because it needed to weigh less than twenty-seven hundred pounds if the Atlas were to launch it into orbit. And through the rest of 1959 there were still problems with weight. The astronaut weighed only 160 or 180, but every pound of man required several pounds for the systems necessary to keep him alive. After accounting for the heat shield, the interior bulkheads, and the parachute system, the engineers had to fit into that tiny space oxygen tanks and backup tanks, a contoured fiberglass couch, an impact bag, a radio, and electrical, hydraulic, and mechanical systems, and backup systems. McDonnell had nine major subcontractors and more than six hundred firms supplying parts, and although the subcontractors and firms miniaturized those parts, even using laboratory scales they consistently underestimated their weights, and so the weight of the whole capsule kept creeping up. By November 1960 the capsule would be too heavy for the Atlas launch vehicle that was supposed to lift it. This presented a real problem, especially because the Atlas's designers, in a weight-saving measure of their own, had given the missile an extremely thin skin. Empty and laid on its side, an Atlas would actually collapse of its own weight, and in one test flight the twenty-seven hundred pounds of Mercury spacecraft, under increased g-loads, had wrenched it like a tin can.

More than a year earlier, in March 1959, the Space Task Group met at the McDonnell plant in St. Louis to examine the first full-scale mock-up of the capsule. Lovelace was there, too. Even from the mock-up, two things were glaringly obvious. One, the automatic systems made it possible for any healthy person to survive the flight. Two, the engineers' work would be easier if they could use a smaller, lighter occupant.

21

JERRIE

In September 1959 Jerrie Cobb was five feet seven inches tall, and weighed 124 pounds. She was twenty-eight-years old, a pilot and manager for Aero Design and Engineering Company, and one of the few women executives in American aviation. Jerrie had logged 7,000 hours of flight time, won three world records, and been awarded the Fédération Aéronautique Internationale's Gold Wings of Achievement. In recent months much of Jerrie's work involved traveling to airshows and conventions, putting the Aero Commander through its paces for potential buyers. One such convention, a meeting of the Air Force Association, took place in Miami Beach that fall. She and her boss, Tom Harris, had flown in to attend.

Early on the second day of the conference, Jerrie and Harris were walking along the beach, discussing their plans for a session that afternoon. It was not yet seven. The sun was barely above the horizon and the beach was mostly empty. They noticed two figures against the glare, men wading into shore. Harris told Jerrie he knew them. When they were near enough, he made the introductions: "Dr. Lovelace, General Flickinger, I'd like you to meet Miss Jerrie Cobb."

Jerrie knew these names. She knew Lovelace ran the clinic where the Mercury astronauts had undergone medical testing. Flickinger's name had appeared less often in the papers, but she recognized it as belonging to one of the most influential figures in aerospace medicine.

Lovelace and Flickinger had just flown in from Moscow, where they had attended a meeting of space scientists. Now they began to discuss Soviet aeronautics with Harris. The four of them walked along the shore together. Unprompted, Jerrie remarked on some of the problems one of the Russian planes posed for pilots. Lovelace stopped walking and looked at Harris's assistant in astonishment. "Are you a pilot?"

When Jerrie admitted to having flown for sixteen years, the physicians were dumbfounded. Surely she wasn't old enough. And Harris said, "Jerrie has more than seven thousand hours in her log. She's set three speed and altitude records for us, and the FAI just awarded her its Gold Wings of Achievement." There was a small silence. Flickinger remarked that it was astonishing what women were doing in aviation these days. Harris, of course, was quite aware of what women were doing, and what they wanted to do. Only a year before he had listened to Jerrie propose to take an Aero Commander on a solo circumnavigation of the globe via the poles. Now he half jokingly suggested that NASA might want to keep an eye on this woman aviator saying, "Jerrie's liable to try for a record in space next."

But Lovelace did not laugh. Instead, he said, "As a matter of fact, we had indications at the Moscow meeting that the Russians plan to put women on space flights." This moved the conversation toward a new and vast subject, but everyone had appointments. Before they parted ways, Lovelace looked at Jerrie seriously and said, "Let's get together and talk some more."

Lovelace reserved a meeting room in the Fountainbleau Hotel. Flickinger and Harris and Jerrie were seated in chairs arranged in a small circle, and after a few moments of small talk, Lovelace looked at Jerrie and said, "Would you be willing to be a test subject for the first research on women as astronauts?" Jerrie was caught short, but Lovelace smiled and continued as though he had rehearsed. "The most recent studies of women pilots under stress were done during World War II, when the WASP flew for the Air Force. Medical studies demonstrated that women in general are better than men at withstanding pain, heat, cold, loneliness, and monotony—all of which are likely to factor into space flight. But none of these tests specifically measured women's capacity as potential astronauts." He continued, saying that although the tests he was planning would not be conducted under the auspices of NASA, they would be the same tests that had been administered to fifty-six men who were called astronaut candidates.

To Jerrie it was all happening so quickly. He had said "women's capacity as potential *astronauts*." She rationalized, told herself that their interest was merely an indication that they thought she could help them experiment. But she knew that the *timing* of this experiment suggested that Lovelace and Flickinger were nursing an idea that went beyond lab tests and scientific papers. America was in desperate need of a first. And as Jerrie was sitting in that hotel lobby, she began to think herself justified in imagining that within this proposal lay the possibility that she, or someone like her, could be that first. The first woman in space.

For six months after her meeting with Lovelace and Flickinger, Jerrie Cobb imagined outer space as any pilot would. She did not think about getting there first or about knocking the Russians off their perch or about becoming a hero. She just thought about what it would look like. Being alone, a hundred miles up, and looking down. It was the most wonderful thing she could imagine.

Formal notification from Lovelace came in December. Jerrie was asked to report to the Lovelace Clinic in February. With the notification came advice: start training—running and swimming and cycling and eating lots of protein but not too much fat, and getting at least eight hours of sleep a night.

A year earlier, when NASA had made the call for astronaut candidates, the services realized the competition into outer space was not over. The program was civilian, but the astronaut himself would be Navy, Marine Corps, or Air Force. So the Navy and the Air Force had set up little etiquette lessons for the candidates who would represent them. At any social function a candidate was to hold his drinks to two, and if he placed a hand on his hip, he was to put his fingers toward the front, thumb toward the rear, and so on. The men had nurtured a healthy skepticism about the etiquette lessons, and the subsequent poking and prodding at the clinic, but Jerrie felt differently. The men's success or failure had been theirs alone, or at best, the success or failure of their branch of the service. But in a way she found hard to explain, Jerrie began to think her failure would be a failure for all women.

On the morning of February 14, 1960, she reported to the Lovelace Clinic to begin the tests. It happened to be Valentine's Day, but there was little romantic about her activities that morning, or for that matter, the following week. She was "unit one, female." Like the Mercury astronaut candidates, she was a sort of specimen. There were examinations by an ophthalmologist, an otolaryngologist, a cardiologist, and a neurologist. There was also an examination by Randy Lovelace himself. She was prodded and poked and scrutinized, and prodded and poked some more. They took X rays, they made films of her heart, they checked her vision and her hearing, they induced vertigo to see how quickly she recovered from it, they gave her electric shocks and determined her lean body mass. At the end of the week she was summoned to the office of Dr. R. R. Secrest, the flight surgeon who had made the initial physical examinations. He smiled and said, "Let me sum this up quickly, Miss Cobb. You are a remarkable physical specimen. You've passed the Mercury astronaut tests."

Randy Lovelace was encouraged by the results, but this was just the beginning. He was calling it Phase I. Phases II and III of the tests—the psychological stress studies and flight simulations—would take place over the next several months and in different parts of the country. Jerrie had done well enough already to warrant further study. At least twelve more female pilots would now be tested. Meanwhile, Lovelace asked Jerrie to keep her participation a secret. He expected that press interest in her would be considerable, and the clinic had no facilities to deal with it. If the reaction to the Mercury astronauts in the preceding months was any indication, he was probably right.

In August 1959 the astronauts had sold their exclusive "personal stories" to Time-Life Inc., in a contract which ensured the astronauts final editorial control over any article appearing in *Life*.* For Lovelace and Jerrie Cobb too, Time-Life seemed the best means to prevent the tests from becoming sensationalized, and to protect Jerrie's privacy. If she would give *Life* an exclusive, they would hold the story until Lovelace made a formal presentation to the aerospace medical community at the Space and Naval Medicine Congress in Stockholm. Lovelace was scheduled to make the speech on August 18. *Life*'s editors asked Jerrie to come to New York for a press conference that would occur a few days later. After the press conference she would be available to reporters from other publications.

Privately, Randy Lovelace hoped that the testing of Jerrie Cobb might lead to something more—an expanded testing program for women, perhaps an American woman in outer space. Publicly, he knew better than to voice such hopes. So at the Space and Naval Medicine Congress in Stockholm on August 18, he made a presentation which was cautious and restrained. He noted that a female subject required less oxygen per minute than the average male, meaning less oxygen by weight would need to be carried for the women crew members than for the men. Furthermore, because women's reproductive organs were internal, there would be less danger from radiation. And women seemed to tolerate heat as well as men. In sum, preliminary results indicated that women would make competent spacefarers, and the tests on women would continue. Lovelace cautioned the audience against assuming that the next step was a woman in outer space. At present, there was no such program. The tests were pure research, justified only by the probability that women would eventually participate in space flight. He headed an advisory committee with no responsibility or accountability, and the tests were administered under his own auspices and with the foundation's money.

On the whole, Lovelace's presentation was fairly understated, and its reception at the Space and Naval Medicine Congress would best be described as polite. The tone of the AP wire written from

*Time-Life paid the astronauts five hundred thousand dollars, an amount to be equally divided among the seven. And since 1959 rare was the issue that did not contain at least a few pages of Astronaut Gordon Cooper carrying groceries into his modest ranch home, or photos of John Glenn's father and mother in their living room in Ohio.

the conference press release was similarly cautious and restrained, observing little more than that a group of medical specialists had performed a series of tests on a female pilot named Jerrie Cobb. Probably the copywriter thought that if newspapers printed it at all, it would be in the back pages.

By 6:00 P.M., EST, on the evening of August 18, the story started coming off the newswires from Stockholm. . . .

That night Jerrie was staying at a friend's apartment in New York. At 3:45 on the morning of August 19, the phone began to ring. Somehow the Associated Press had tracked her down. Her friend took the call and, with a groggy Jerrie Cobb standing nearby in a borrowed bathrobe, she said the test subject was no longer there; she further suggested that she thought Jerrie was driving back to Oklahoma. The phone rang all night long, and the phone rang at Jerrie's parents' house too. The next day calls came to her office in Oklahoma City. Some were from as far away as Australia and Japan.

Because this was a science story the reading public could understand, and because such stories are few, American newspapers reported it. Lovelace was not in NASA's employ, the testing was not NASA's, and there existed no plans to put a woman into space, let alone an astronaut training program for women, but the newspapers regarded all these matters as details so minor as to be almost unworthy of mention. Here it is, they said: Medical science has shown that women are as suited for outer space as men, and in some respects actually better! Further, the word had come from not just any researcher, but from Lovelace himself, one of the most respected figures in the history of aerospace medicine and a specialist highly regarded by NASA. If Randy Lovelace wasn't exactly a NASA spokesman, he was certainly close enough. And if Jerrie Cobb wasn't America's first woman astronaut . . . she was close enough, too.

It was useless to explain to the press that the tests were just tests. And Jerrie couldn't explain anyway because, abiding by the contract with Time-Life, she was in hiding. She spent three days holed up in New York while her friends at Aero Commander tried to convince the newspapers that she was flying to South America. They were not persuaded; one British paper tried to gain access to her by threatening to print a story that she'd been kidnapped by the Russians.

Finally, before a capacity crowd in the Time-Life auditorium, Jerrie Cobb, unit one, female, was at last presented to the press. She had never felt comfortable speaking before a crowd, but she swallowed hard and walked onto the stage. Newsreel and television teams adjusted lights and microphones. Then began the first of a series of interviews. These weren't as difficult as she'd expected—the interviewers seemed to have a talent for putting people at ease. A few hours later they were over, and she faced a more difficult audience, a circle of newspaper and wire service reporters. Some questions she had anticipated. "What advantages would a woman astronaut have?" She offered answers which repeated Lovelace's conclusions—greater resistance to radiation, tolerance for pain, and so on. The reporters scribbled in their little spiral-ring notebooks. But they could have found that in the AP wire, and they knew it. They wanted more from her. Another question: "What do you have to do with proving this?" Her answer: "Since space is new and strange to all of us, including scientists, women must be tested and rechecked in the situations we believe space will demand. There has been no previous testing of women in the light of possible space flight. I was asked to be the first subject." The reporters began to shift in their seats. They had deadlines. One journalist, desperate for a bit of color—something, *anything* . . . —asked, "Can you cook, Miss Cobb?" To this Jerrie smiled, and answered politely. She could cook, and did, when she had time. She was partial to Chickasaw Indian dishes—steamed beef, dried corn, squaw bread, and a water lily root called a yonkapin. They scribbled some more in their little spiral-ring notebooks. At least they were getting closer. Finally, somebody thought of a way to ask it. "Miss Cobb, how do you feel personally about space?" There it was. She looked directly at him. It was a question she'd asked and answered herself in the months of waiting, and the week of testing. She'd suspected the answer all her life and known it with utter certainty from the day she'd flown the Aero Commander to the lower edge of the stratosphere. And now she offered that answer with the ease and quiet conviction with which any priest recites the church's creed. "*I feel that space is God's unspoiled world which humans should not trespass upon without a feeling of reverence.*"

The interviews and the news conference were easier than Jerrie thought they would be, but she was relieved when they were over.

She looked forward to getting home, to a quiet weekend with her parents, to catching up on sleep. But it wasn't over. At the Oklahoma City airport, on the day she arrived from New York, there was a small ceremony, complete with local dignitaries and a reviewing stand. The governor himself was there, and he named Jerrie Cobb Oklahoma's "Ambassador to the Moon." When she arrived at her office, she found piles of letters heaped on her desk, many from people wishing her well, and several telegrams from major corporations who wanted to use her in their advertising campaigns. She was offered a chance to endorse mattresses in magazine ads and to star in a series of television commercials for cigarettes. In the days that followed she turned down all such offers simply because she didn't want to trivialize the Lovelace project. But she made an effort to answer all the letters.

In the fall of 1960 there were rumors that the Russians were about to score another first—perhaps putting a man in space. And at the McDonnell plant there were still problems with the weight of the capsule, so it was hard for Randy Lovelace not to think of a twenty-eight-year-old aviator, weighing in at 124 pounds.

By this time, Lovelace and Flickinger had already planned Phase II of Jerrie's tests. It was designed to replicate part of the stress tests the Mercury astronaut candidates had undergone at Wright Patterson. Some of these tests had evolved from the concerns of a certain sector of the aerospace medical community, specifically, a group of behavioral psychologists. The psychologists thought the talk of pilots and ejection at supersonic speeds was all very interesting, but they were concerned with behavior which was considerably less athletic. They were more interested in men sitting still in little dark climate-controlled rooms called "space simulators." By August 1960 there had already been several series of tests using space simulators—a kind of intra-service race to inner space. In early February 1958, Airman First Class Donald G. Farrell spent a week in a space cabin simulator at Randolph Air Force Base. One month later, five Air Force officers made another simulated space flight for five days. A month after that, at the Philadelphia Naval Base, six Navy men made a simulated seven-day flight to the moon.

In April 1960 a man spent a week in a simulated space capsule environment at the USAF Aerospace Medical Laboratory. And in August of that same year, two pilots at Brooks Air Force base made another simulated flight to the moon, this one for seventeen days.

What a simulated space flight meant, exactly, was that they would shut the door and turn out the lights, and the subject would sit in a reclining chair and watch the simulated control panel, maybe a piece of plywood punched with holes through which were inserted little lights and circuit breakers. He would watch those lights and perform a few simple stimulus-response exercises. Maybe somebody outside would talk with him over the radio. Maybe not. He would eat low-residue food and make sure his waste went into plastic bags. Mostly though, he would just sleep a lot. A week later they would open the door, and he would get up from the chair and step out. He would squint as his eyes adjusted to the bright room lights, and then maybe he would yawn and stretch. And he could say he had been on an actual simulated flight into outer space.

Why sitting in a room for a week had anything to do with actual spaceflight was an interesting question. The answer from certain psychologists was a fear that the subject might grow disoriented in the absence of gravity, or worse, he might panic and start punching buttons that would vent his oxygen into space or make him reenter at too steep an angle. There was just no way of knowing for sure. There were subtler dangers, too. This person would be separated from the rest of humanity in a manner that was utterly without precedent. A psychiatrist by the name of D. G. Starkey offered a dark warning: "Isolation produces an intense desire for extrinsic sensory stimuli and bodily motion, increased suggestibility, impairment of organized thinking, oppression and depression, and in extreme cases, hallucinations, delusions, and confusion." Some had unspoken fears of a recovery crew opening the hatch and finding the passenger with his eyeballs rolled up into his skull and his legs and arms twitching uncontrollably.

At any rate, this conception of spaceflight as an experience of unprecedented isolation encouraged a belief that the astronaut should be a perfect or near-perfect *psychological* specimen. Such concerns were voiced not merely by ivory tower psychologists, but by persons with considerable experience in aircraft. Air Force Colonel John Stapp, director of the aeromedical laboratory at Wright Field, was known to the general public as the man who, in the interests of gaining a knowledge of acceleration on the human body, had survived tremendous g-forces by riding a rocket sled that accelerated from zero to 632 miles per hour in a matter of seconds. The same Colonel Stapp had also managed a number of isolation tests; in 1959 he had announced the conclusion of a study which found that in cramped quarters, women are more likely to remain psychologically stable. In fact, one woman—her name did not appear in the press releases, and she was not associated with Lovelace's tests—had lasted five days in an Air Force test chamber. The previous record, set by a man, had been one.

In September 1960, Jerrie Cobb underwent Phase II of the astronaut tests. As it happened, she would not travel far. The tests would be administered right in Oklahoma City, in the laboratories of the psychiatric services division of the Veterans Administration Hospital. There were electroencephalograms, and readings of her brain waves while she slept. There was a battery of psychological examinations—Rorschach tests, the Wechsler adult intelligence scale, sentence completion tests, "draw a person" tests, and the Minnesota Multiphasic Personality Inventory. On the last day, there was a sensory deprivation test.

Dr. Jay T. Shurley had begun work in sensory isolation at the National Institutes of Health in Washington. At the Veterans Administration Hospital, he had designed the most comprehensive sensory deprivation environment in existence. It was a pool of water kept at skin temperature, inside a room made soundproof, vibration-proof, and perfectly dark. The forty or so test runs so far had yielded a variety of responses. One man was in the tank for a few hours before he began singing obscene songs, crying for all the people who "never stop to think of what life is about," and insisting that "the voices" be quiet. After four and one-half hours of this they turned on the lights and yanked him out.

To Jerrie, the isolation tank was infinitely preferable to the press conference. She regarded it as a pleasant, if unusual, vacation—a quiet, warm bath for the length of a workday. Floating in the darkened pool she planned the weeks ahead. When that was done, she worked out a crick in her back, then just relaxed. She was in the pool for nine hours before the psychologists decided there was no point in making her stay any longer. In fact, her nine hours was the second longest anyone had floated there. A few days later, Shurley's report summarized her performance:

> Thus far in our experience, probably not one in 1000 persons would be capable of making such a lengthy isolation run. . . . She would be in the top 1% or 2% to adapt. . . . Based on the above partial assessment through psychiatric and psychological techniques and augmented by a special test under conditions of profound experimental sensory deprivation, it is our opinion that Miss Jerrie Cobb . . . possesses several exceptional, if not unique, qualities and capabilities for serving on special missions in astronautics. . . . Among these are: A ready acceptance of direction or a ready assumption of responsibility, as circumstances dictate. An exceptional ability to remain passive and relaxed when action is unavailable and unwise. An unusually smooth integration of psychophysiological functioning, a stable ego, and a strong, healthy motivation. Her pleasant personality would lend itself as well to nonsolitary missions. I believe she has very much to recommend her as an astronaut candidate.

Actually, exactly how much Jerrie had to recommend her as an astronaut candidate depended on how one defined an astronaut. Within NASA, even by late 1960, the matter was far from settled. And the value of Shurley's tests were also much contested. Some members of the psychiatric community were telling NASA that although a psychologically healthy person was preferable, the isolation tank stuff seemed a bit much. The Mercury spacecraft, after all, was not an isolation tank.

The pilot/Spam matter had been much discussed within the Space Task Group, the engineers at McDonnell, and among the seven astronauts themselves. But to everyone else, the reasons for concern about the psychological health of the candidates and the degree to which the astronaut would pilot the spacecraft were difficult to understand. As for the journalists, they knew of the pilot/Spam debate, and understood that it was often framed in such a way as to portray pilots as superior to Spam. But either they didn't understand exactly *why* pilots were portrayed as superior, or they didn't know how to explain that superiority to their readers. So they simply wrapped the pilot-astronaut and his cherished control in the American flag. The Russians controlled their manned spacecraft exactly as if they were unmanned, from the ground. But NASA's approach was true to the spirit of a free society: Americans in space, controlling their own fate—or at least their roll, yaw, and pitch, at least some of the time. The public, insofar as it understood this distinction, approved.

For her part, Jerrie began to realize that the job description for astronaut varied depending upon who you asked, when you asked, and where you asked it. In fact, there seemed to be two job descriptions. One described an active pilot, a thinking, cogitating, controller. The other described an inert medical subject, the perfect physical specimen, the . . . *Spam*.

In the summer of 1960 Jerrie had flown an Aero Commander across the South Atlantic Ocean, then throughout much of Africa—demonstrating the plane in Monrovia, Praetoria, Nairobi, Tanganyika, Zanzibar, and Johannesburg. That same summer NASA's Lewis Research Center in Cleveland allowed her to undergo part of a simulated spaceflight in a machine known as the Multi-Axis Spin Test Inertia Facility (MASTIF). It was a great two-story rig of three frames of steel tubing, one nested within the other on gear wheels and ball joints so that each frame could spin independently. Inside the central frame was a pilot's seat. The MASTIF in action was supposed to simulate a spacecraft tumbling wildly out of control, and the test subject's job was to use a control stick to stabilize it. By 1960 it was part of the astronaut's training, and it was difficult; on his first run Alan Shepard had hit the chicken switch and ended the simulation. So when Jerrie was strapped into the pilot's seat the technicians expected she would end the test quickly. With a grinding sound, the outside frame began spinning slowly at first, like a record on a turntable, then faster and faster. Then, while the first frame did not slow its rotation at all, the second frame suddenly pitched forward and over and over. Finally, the inner frame began to spin her clockwise like a sock in an automatic laundry dryer. Now the MASTIF was a blur of screeching metal. Through sheer force of will Jerrie fought increasing nausea, and fixed her eyes on the instrument panel. Then, remembering the controls were exquisitely sensitive, she gradually slowed and stopped motion on one axis, and then another, then the last, until the great contraption was quieted and still. The technicians helping her out of the harness were impressed.

A few months later, in September, Jerrie flew a stripped Aero Commander 680F to 36,932 feet—a world-class altitude record for light aircraft. Her record stay in the isolation tank in Oklahoma City was accomplished in the same week. All that long summer and fall, it was as though she were saying to NASA, *You want a pilot? Then I'll be a pilot. You want Spam? Then I'll be the best Spam in Spam history*.

When the seven Mercury astronauts were presented in 1959, the public responded with an outpouring of adulation. Here were seven men who looked like normal men and yet were not normal men. They were astronauts. They were—the future. But not to everyone. Chuck Yeager himself, arguably the best pilot in the United States, had been ineligible for Project Mercury because he had no college degree. He didn't seem to mind. When a reporter asked him if he was interested in Project Mercury he chuckled a little and said, "No, it doesn't really require a pilot, and besides, you'd have to sweep the monkey shit off the seat before you could sit down." Yeager's career was in no danger of faltering because NASA didn't require his services. Since late 1959 Yeager had been squadron commander at George Air Force Base, about fifty miles from Edwards, where he was working on the operational deployment of the F-100 Super Sabre.

George Air Force Base wasn't far from the Cochran-Odlum ranch, and Chuck and Glennis still saw a lot of Jackie and Floyd. In fact, Chuck and Jackie had only recently returned from a two-month trip to the Soviet Union. The opportunity had arisen because Jackie had been elected the first woman president of the FAI, and the organization was holding its annual meeting in Moscow. She planned to fly herself in her twin-engine Lockheed Lodestar, and she wanted Yeager along as copilot and navigator. She approached the Edwards chief of staff (Yeager was working with the X-series planes at Edwards then), and he approved the plan, as long as Yeager would travel in civilian clothes with a civilian passport, and keep his eyes and ears open for information regarding Russian aeronautical activities which might interest U.S. intelligence.

In the early summer of 1959 the plane took off from LaGuardia carrying Jackie and Yeager, Jackie's private secretary, her hairdresser, and a woman from the State Department who would act as her Russian interpreter. The baggage compartment was so crammed with trunks of clothes that Yeager joked they would be lucky to get off the runway. It was an eventful trip. Yeager befriended several of the Soviet Union's test pilots without surrendering any secrets, took a great many photographs of MiG bases, and flew the Lockheed Lodestar at 10,000 feet over miles of Arctic icecap. For her part, Jackie made a landing in zero visibility and a forty-mile-an-hour crosswind on a grass strip runway in Kiev, and entertained delegations from dozens of countries in her Moscow hotel suite. Jackie was frustrated with Soviet inefficiency and regulations, and made no secret of it. Near the end of the trip, she shouted down a Red Army general, prompting Yeager to wonder aloud whether she was risking a stay in a gulag. It was one of numerous occasions on which she had tested even Yeager's considerable patience. When they returned, he told Glennis he'd never travel with Jackie again. But Glennis just smiled a little and reminded him that he always said that after a trip with Jackie.

As for Project Mercury, Chuck rarely mentioned it. Jackie Cochran didn't mention it either. Until she learned that Randy Lovelace was running his astronaut tests on women.

Lovelace had known Jackie and Floyd since before the war. Jackie had tested prototypes of the oxygen mask which Lovelace and his associates were developing in the 1940s. In 1937 Lovelace was among the team of physicians who, in part through Cochran's efforts, had been awarded the Collier Trophy for research in problems associated with high-altitude flying. And in 1947, when Lovelace established the foundation, he asked Floyd to act as its chairman, and Floyd generously funded a good deal of its operations. By 1960 Lovelace was the personal physician of both Jackie and Floyd. In the fall of that year he mentioned to Jackie that it might be worth running some tests to see how women would fare as astronauts. Lovelace said he thought she might help him find suitable candidates.

Jackie had hopes that she might be one of those candidates. But she had serious strikes against her. Her health was generally poor; she had suffered a botched surgery for appendicitis when she was a child, and experienced persistent abdominal difficulties that had required a series of operations and resulted in two miscarriages. Further, she was in her fifties. But she had one point in her favor: Randy Lovelace. Generally, test pilots regarded physicians as hindrances to aerospace advancement and a visit to a flight surgeon as something to be avoided. But she knew that Lovelace had spent much of his career discovering ways to allow pilots to fly. An X-15 pilot named Scott Crossfield had suffered sinus and ear damage from pressure changes; corrective surgery by Lovelace had kept him in cockpits. Aircraft mechanics repaired planes; Randy Lovelace repaired human beings. Why couldn't the doctor make a few . . . *fixes*?

Jackie told Lovelace she'd be happy to do what he asked, but suggested she could do even more. While Lovelace would perform the tests at the foundation's expense, his plan was for the women to pay for their own travel and lodging. Jackie promised not only to find the women but to cover their personal expenses. She and Lovelace combed the rosters of the Ninety-Nines. Jerrie Cobb, meanwhile, submitted a list of women pilots she thought suitable, and they went over that too. In the end, they chose twelve women, and then drafted a letter. For Jackie it was almost like old times again, like the early days of the WASP. Only this time she couldn't help feeling that she was being asked to stand on the sidelines and just watch.

On January 25, 1961, the story was released over the AP wires. The headline read "12 WOMEN TO TAKE ASTRONAUT TESTS." Those who were interested enough to read on would learn that, over the next few months, at least a dozen women pilots would travel to the Lovelace Clinic in pairs to begin the testing process through which the original astronauts had been selected. But in part because no names were released, this announcement did not cause the frenzy that Lovelace's report on Jerrie Cobb had unleashed a few months earlier. By January 1961 the public's attention was focused on the seven men who were preparing to fly into outer space. Under the circumstances, the report that the Lovelace Clinic was continuing its experiments on women went virtually unnoticed.

22

LOVELACE CLINIC

Jerri Hamilton was born in 1931 in Amarillo, Texas, in the middle of the Panhandle. A few years before, a great oil field had been discovered up north, and after that it seemed everybody had become a driller or a foreman. Like most men in Amarillo, Jerri's father worked for the oil companies. Some days, when the wind picked up, it would kick up so much dust that the whole sky would darken. As a matter of course drilling companies supplied their workers with masks made of rubber and canvas, with a little round piece over the mouth to unscrew when the filter needed to be changed. They gave extra masks to the workers to give to their families. And the children of Amarillo would carry their books to school, leaning into the wind, wearing dust masks.

On Saturdays, whether the wind was blowing or not, the children of Amarillo went to the local theater for a movie starring Gene Autry or Roy Rogers. Little Jerri went too. Most of the kids there wanted to be cowboys, and some of them were already wearing five-gallon hats right there in the line for popcorn. Jerri liked Roy and Dale enough, but mostly she was there for the trailer, a serial about a pilot named Tailspin Tommy. Tailspin Tommy had a girlfriend, and she was a pilot too. Jerri knew she wanted to be a pilot like Tommy or his girlfriend. As a child Jerri began to watch planes, and the sky. By the time she was in third grade, she already had her own leather helmet and goggles. They were practical there in the Panhandle, especially during the winter—the helmet was warm and the goggles kept the wind out of her eyes. Her mother liked to curl Jerri's hair before school into fifty-two little Shirley Temple curls. Jerri would submit to this ritual patiently, but when she got out the door she'd cram that helmet down right over those curls.

By the time war had come Jerri's parents had separated, and Amarillo itself had changed. In the 1940s, a lot of young men from Amarillo had gone to the Army or Navy, and so it was a town populated by women and girls, recent veterans, and men too old to fight. It was obvious to anyone that a war was on, especially when a train came in. Amarillo was where the Burlington Line, running north-south, crossed the Santa Fe Line, running east-west. Both railroads had depots downtown. From the Burlington Depot, one could walk south a few blocks, take a left at Fox Drugs and then walk two more blocks to the Santa Fe, where one could board a train for Albuquerque and points west. Fox Drugs was owned and managed by Jerri's stepfather.

One evening in 1944 when Jerri was thirteen, she and her mother took the car downtown to meet him. Because gas was rationed, even families with two cars ran only one. In the fading evening light Jerri and her mother sat in the car, parked across the street, waiting for Pop to turn out the lights and lock up. They talked a little and watched the soldiers walking from one depot to the other. Suddenly, around the corner came two women, walking along with the soldiers. They were wearing handsome blue uniforms, and carrying B-4 bags over their shoulders. They seemed to be changing trains too. Sitting in the car with her mother, Jerri leaned forward until her face was touching the glass, and made that astonished pronouncement, "They're lettin' women in the Air Force." Her mother said, "Those are WASP. They're in the Ferry Command." For Jerri, such fine bureaucratic distinctions were beside the point. Women were flying airplanes for the United States Army Air Force. At that moment Jerri decided she would join the WASP.

A few years after that night in front of the drugstore Jerri took her allowance and paid a man at Tradewinds Airport at the edge of town to teach her to fly a Waco. The problem was that the allowance her mother gave her was meant for gas and lunch money, and flying lessons could get expensive. Jerri had a car, and since several friends were riding with her to school, she began to charge them for gas and wear and tear. In fact, they were funding her flying lessons. This went on for a year before her mother found out. When she did, all heck broke loose. Mrs. Hamilton believed flying lessons were improper for a young lady of Jerri's social standing, and thought Jerri needed to be reined in. She was sent south for a year, to Saint Mary's, a Catholic high school in San Antonio. For that whole long year she listened to the sisters and learned the catechism. Mostly though, Jerri just plotted how she might get back into an airplane cockpit.

The next year she finished at Amarillo High, and the WASP had already been disbanded. Soon after graduation she met a man named Louis Sloan, and Jerri and Louis found they had much in common. He'd been a B-24 pilot in the war, and like her, he wanted to be near airplanes. Soon he and Jerri were married. On Louis's GI Loan, they both enrolled at the University of Arkansas, mostly because it had a flying school. They were both working and attending classes, but soon enough there were children—a girl they named Candy and, a few years later, a boy they named David.

Jerri wanted a career as a commercial pilot, but a commercial license required 200 hours, ground school, and an instrument rating. *Two hundred hours.* To someone whose work is piloting, 200 hours just added up when he wasn't looking, but for anyone else, 200 hours was almost impossible. It was hard for a woman to gain enough flying time to get a commercial license, so Jerri began to enter air races. The hours began to add up, and soon she gained her airline transport rating, her flight instructor rating, and her instrument rating. She also won quite a few races. Jerri Sloan had met Jerrie Cobb in one of those races, and after the first jokes about their names, they became fast friends. Jerrie Cobb had an integrity and decency that Jerri Sloan admired, and Jerri Sloan had an articulate manner that Jerrie Cobb envied.

At Addison Airport in Dallas, Jerri Cobb learned one day that Texas Instruments was about to release an RFP for airborne testing of electronic surveillance equipment. She knew that as a woman she'd have little chance to get the contract, but she had a friend there. His name was Joe Truhill, and she asked him if he'd like to pool resources. Jerri and Joe began an aircraft company called Air Services at Addison Airport, and they won the contract with Texas Instruments. She couldn't believe her good fortune. For the first time Jerri was making good money flying an airplane. The work was hard and the days were long, but they were seeing results. By 1959 they had a B-24, a B-25, and a DC-3. They also had a steady stream of contracts from TI. Moreover, the work was interesting: That initial contract had been for testing the first infrared imaging system in the world.

The Texas Instruments hangar was next door to the Air Services hangar, and Jerri was Air Services' chief security officer. These were people who joked about a lot of things, but when it came to security, they were stone serious. Recently, some of their work played a part on the international scene. The infrared system they had flown had been aboard the U-2 piloted by Francis Gary Powers, shot down over the Soviet Union in May. (As it happened, the same August day that the AP story on Lovelace's initial tests on Jerrie had come over the wires, Powers pleaded guilty of espionage before a high military tribunal in Moscow.)

One day in Dallas, Jerri received a phone call from Jerrie Cobb. Jerrie asked her if she might be able to get away for a few days to work on a secret government program. It sounded interesting—Jerri Sloan had already been working on secret government programs, so she didn't hesitate before she said yes. In such matters, one was given information on a "need-to-know" basis, and Jerri assumed that Jerrie didn't describe the project because she didn't need to know, so she didn't ask. She hung up the phone. There was a lot to do that year. She and Louis had separated, she was raising Candy and David on her own, and every week the Texas Instruments engineers seemed to show up with something new. She didn't have time to think much about that phone call until a year later, when the letter came.

Dear Miss Sloan:

We have been informed that you may be interested in volunteering for the initial examinations for female astronaut candidates. These examination procedures take approximately one week and are done on a purely voluntary basis. They do not commit you to any further part in the Woman in Space program unless you so desire.

There is no charge for the examinations. Your only expenses would be traveling expenses and room and board in a nearby motel. Enclosed is a card which outlines qualifications of women astronauts. If you would prepare a curriculum vitae following the outline and return it to me, we would be happy to consider your application. Under separate cover we are sending a brochure which describes our foundation and clinical organization.

We shall look forward to hearing from you if you are interested in this program.

W. Randolph Lovelace II, M.D.
Director

Female astronaut candidates? Woman in Space program? Qualifications of women astronauts? Then she remembered, and looked at the name under the signature again. It was Lovelace. She knew he'd been the director of the medical examinations for the astronauts; it had been in all the papers. She had opened the letter in the kitchen, and David had run in from outside. She smiled and said, "Mom's going for astronaut testing." She explained it a little, then he ran outside to tell his friends that his mother was going to the moon. As the screen door banged she repeated to herself that last line above the signature. *If you are interested.* . . . Jerri Sloan was more than interested.

Other women were recruited, twenty-six in all. They were established pilots in good physical condition. Each was asked to undergo tests at Lovelace. The tests themselves went on for ten days. From January to July, the whole first half of 1961, the women were put through it two at a time.

Each test subject went to Albuquerque, and checked into a room at the Thunderbird Hotel across the street from the clinic. The evening before the tests there was a knock on the back door. A man was there asking her to sign papers absolving the clinic of legal responsibility. On the next morning she woke at 5:00 A.M., showered, dressed, and with little or no breakfast as per instructions, walked across the street to the Lassetter Laboratory Building, through the glass doors and into the lobby. There she presented herself to receptionist Vivian Thomas. Miss Thomas smiled politely, assigned the test subject a number, handed her two enema bottles and a five-page schedule of tests, and gave her directions to Room 2B, where it began.

The eye examination included refraction, visual fields, extraocular muscle balance, red lens test, tonometry, depth perception, slit lamp, dark adaptation, and dynamic visual acuity . . . The otolaryngolocal tests included visual inspection, indirect laryngoscopy and nasopharyngoscopy, audiometric thresholds, speech discrimination, and labyrinth function by the standard caloric method . . .

Examination by cardiologist included electrocardiograms and ballistocardiograms. A tilt-table test was done, in conjunction with the physiology section, to acquire information on the stability of the pressor-reflex mechanisms and the effectiveness of vasomotor control by the automatic nervous system . . . The Lee and Gimlette procedure was employed by an expert to detect congenital abnormal openings between the right and left sides of the heart . . .

Myrtle Cagle was thirty-six, a flight instructor in Macon, Georgia. Her friends called her Kay. She was a small woman, five feet two inches and only 110 pounds. She had 4,300 hours, and ratings as a flight instructor, instrument instructor and ground instructor. As a child growing up in North Carolina she had jumped from a second-story window with only a pillowcase to slow her descent. Her brother taught her to fly a plane when she was twelve, and five years later her mother demanded of the high school principal that she be allowed into the high school's new aeronautics class, saying, "She's already a pilot, for gosh sakes." Kay was admitted into the class, and later in the term when the instructor was drafted, she took over his duties. Soon she had an airplane mechanic's certificate and a nursing license. She also had a husband named Walt. In 1960, when she saw a piece in the paper about tests on women, and wrote to Lovelace Clinic, Walt was behind her all the way.

The neurological examination included testing the reflexes and coordination, determining the normalcy of cerebellar function, and determining proprioception and other senses. Dr. L. D. Amick ascertained the conduction velocity of the right ulnar nerve between the elbow and the wrist.

Irene Leverton was single, and thirty-four. She had dark curly hair and compact features. She would say that she took after her German mother, by which she meant that she was strong-willed. On Saturdays when Irene was growing up in Chicago she and her mother would go to Riverview Park just for the parachute rides. Irene wanted to fly P-47s, and when she was seventeen she tried to join the WASP using a faked logbook and an older friend's birth certificate. It didn't work, but the setback was temporary. By 1961 she was the supervisor of a California flying school, with ratings in multiengine land and seaplanes and transports; and more than 9,000 flying hours. Irene had received a call from Lovelace himself; he did not call it a "Woman in Space" program, but when she was at the clinic some of the technicians were suggesting as much. Irene was by disposition a realist, so although she had hope that the tests would lead to something more—a training program, perhaps even an American woman in space—she did not allow herself to think about it much.

An electroencephalogram was done, including a determination of the effects of hyperventilation . . . Proctosigmoidoscopy was performed by a surgeon.

Jan and Marion Deitrich were identical twins, thirty-four years old. They were beautiful brunettes with large eyes and wide smiles. People said they looked a little like Jackie Kennedy and a little like Natalie Wood. Both had graduated from the University of California at Berkeley in 1949. Jan was a quiet woman who had been chief pilot for a flight school and then company pilot for a California construction firm. By 1961 she had a phenomenal 8,000 hours of flying time, and had become one of the first women in the country to earn an airline transport pilot's license, qualifying her to pilot a commercial airliner. Her sister Marion was a general reporter and feature writer for the *Oakland Tribune*, and her work always seemed to lead to airplanes. For one assignment she had flown as a passenger in a supersonic jet. On weekends and vacations she had flown enough charter flights and ferried enough Cessnas and Piper Cubs to have logged over 1,500 hours. Marion, who had received a one-year scholarship to the San Francisco Conservatory of Music, was regarded as the more "poetic" of the two. There was a little sulfur pond near the airstrip where she had learned to fly, and on clear summer nights after work, Marion used to float on her back, looking up at the stars, imagining she was in outer space.

When the letters came Jan and Marion couldn't wire back fast enough—"It would be a great honor to participate in the preliminary or any other tests for woman astronaut candidate."

Laboratory tests under the direction of Drs. T. L. Chiffelle and P. V. Van Schoonhoven [including] complete blood count and special hematology smear, hemoglobin, hematocrit, sedimentation rate, fasting blood sugar, cholesterol, blood grouping, sodium, potassium, carbon dioxide, chloride, urea clearance in blood and urine, blood urea nitrogen, catecholamine, protein-bound iodine . . .

Charles Donlan of the Space Task Group had said for the record that the tests given the men were conducted "as much for the research value in trying to formulate the characteristics of astronauts as for determining any deficiencies of the group being examined." In other words, they were tests of tests. The process itself seemed to inspire a kind of reverence. One late January evening, after a day of tests, Jan Deitrich wrote to her sister, already scheduled to follow her in a few weeks, "Under normal circumstances pain often means injury—and this is probably what is upsetting to most people. But at Lovelace at no time do they do anything that could possibly injure you . . ."

protein electrophoresis, blood volume (Sjöstrand's carbon monoxide method), total body water determination by the tritium dilution method of Pinson and Langham . . .

Jean Hixson was thirty-seven, and a former WASP. She had graduated from Avenger Field with the class of 44-6 in August 1944, and for the last weeks of the program flew B-25s as an engineering test pilot and took an advanced instrument course. After the war she became a flight instructor at the Akron Municipal Airport. Evenings and summers she attended Akron University, where she earned a degree in elementary and secondary education, and in 1952 she began a career as a teacher in the Akron school system. She developed an astronomy curriculum, and by 1961 she was leading field trips to NASA's Lewis Research Center in Cleveland. Jean thought those field trips were as near to outer space as she'd ever get. Then she, too, got a letter from Lovelace.

> *. . . bromsulphalein-dye liver function test, gastric analysis, urine analysis including colorimetric determination of 17–ketosteroids, throat cultures, and stool examination . . . The amount of potassium 40 was determined in the whole body counter at Los Alamos by Langham and Anderson . . .*

To undergo the tests in Albuquerque, Jerri Sloan had to lose a week of work at Air Services and ask her mother to watch the children. Each evening after a long day of needles and tubes and assorted bodily violations, a very tired test subject Sloan walked across the street with her plastic gallon jug. One evening the hotel manager was standing there in the open door of his office. He knew what was going on, and in fact he had told Marion Deitrich that his pep talks had kept other candidates from leaving midway through the tests. He'd heard about all the tests, including the series of self-administered barium enemas. That evening, as Jerri turned the key to her room, he laughed and said, "I'll just bet you're the cleanest woman in the whole state of New Mexico." She narrowed her eyes and said, "When this is over . . ."

Jerri had been paired with a thirty-five-year-old woman named Bernice Steadman. Jerri had heard of Bernice from air races, but Jerri was mostly in Texas and Bernice mostly in Michigan, so they had never met. Bernice owned and operated a flying school at the Flint, Michigan Airport. She had grown up near Detroit, not far from Romulus Air Force Base. As a girl she took flying lessons and planned to become a WASP. When the program ended before her eighteenth birthday, Bernice and her best friend crossed the border into Canada, found a recruiting office, and applied to fly for the RAF. A few weeks later, back home, they received letters notifying them that His Majesty did not require their services. Years passed, and employment prospects did not improve, so Bernice decided that she would have to hire herself. She began her own regional charter and transport service. She also gave flight instruction, and married a man who had been one of her students.

At night, Jerri and Bernice sat on lawn chairs on the little cement porch of the motel. Because they had to save their urine for presentation to the lab technicians in the morning, they had their gallon plastic jugs with them. They traded "war stories" about the air races, and they laughed thinking that what had been performed on them had also been performed on the Mercury astronauts—America's heroes in hospital smocks, lying on their stomachs on cold aluminum tables, and suddenly finding themselves at the business end of barium enema number seventeen.

The next morning the doctor adjusted the headpiece on the chair for Jerri and said, "Now lean all the way back, make yourself comfortable, and watch the light." Jerri looked at the light and said, "What's this one about, Doc?" He didn't look up from whatever he was doing, and muttered, "We are going to inject supercooled ten degree liquid water into your inner ear for thirty seconds. The water will freeze the inner ear bone, destroying your sense of balance, and bring about mystagmus, or vertigo." And Jerri just said, "Oh." She knew there was more to this than just getting through it. She thought the doctors and lab techs were looking for an excuse to flunk her. Any excuse would do, and Jerri was not about to supply them with one. She knew she would have to get through it, and get through it without flinching. Just another little game, another day exploring the limits of human physiology. When she felt the water hit her inner ear and saw the light split into five or six lights which began to spin, faster and faster, she heard a nurse counting seconds and the doctor saying, "Eyeball oscillations, pupils pinpointed . . ." A minute later the doctor said, "We're ready for the other ear." Jerri managed a weak smile and said, "You . . . don't . . . like me . . . do you?"

And the tests continued . . .

In the radiographic examinations, appreciable reduction in radiation exposure was accomplished by the use of supersensitive intensifying screens and shielding plus the use of ultrafast X-ray film.

Under the direction of Dr. J. W. Grossman, roentgenograms were made of the teeth, the sinuses, the thorax posteriorly-anteriorly in inspiration and expiration, and right laterally (searching especially for bullae), the esophagus, the stomach, the colon, and the lumbosacral spine, and cineradiograms were made of the heart (searching for preclinical evidence of arteriosclerosis).

Physical competence tests were administered by Dr. U. C. Luft to provide an estimate of the candidate's general physical condition and cardiopulmonary competence. Graded work was done on a bicycle ergometer, increasing the load from 300 mkg/min to around 1200 mkg/min under electrocardiographic monitoring for possible abnormalities at maximum effort. The heart rate, blood pressure, respiratory volume, and respiratory gas exchange were measured each minute. . . .

Sarah Lee Gorelick had been working as an electrical engineer for AT&T in Kansas City. She received a telephone call from Lovelace on a Saturday in April, and by Monday morning she was in Albuquerque. Sarah had a B.S. in mathematics with a minor in physics and chemistry. She was twenty-eight, single, a commercial pilot qualified in gliders, multiengine and single-engine seaplanes, and had flight instructor ratings. For all her accomplishments, Sarah had a great sense of fun. Once she had taken some girlfriends on a trip to Mexico, and had flown an unpressurized plane over mountains at altitudes in excess of 18,000 feet. Her nonpilot friends complained of headaches, and Sarah said it was because they had skipped breakfast. For Sarah the tests at Lovelace were not so much a ritual to be endured, as they were part of a grand adventure. Every morning she would walk into the offices of that building and present herself with bright eyes that seemed always to say, "What are we going to do today?"

Measurements were made of the total lung capacity and its various subdivisions by direct and indirect spirometry, and the efficiency of ventilation was determined by continuous recording of the dilution of nitrogen while the subject breathed 100 percent oxygen. The timed vital capacity, maximal breathing capacity, and ventilatory response to light exercise (walking at 2 mph for 3 minutes) were determined . . .

Since that AP story in January, the nation at large had heard almost nothing of these women. In fact, the nation at large had other concerns. In January a chimpanzee named Ham was launched on a suborbital flight over the Atlantic. There were some jokes about that name. It was derived from "Holloman Aerospace Medical Center," where he had undergone conditioning, but it was hard not to notice that it rhymed with "Spam." Nonetheless, it was obvious to anyone that America's space program was moving forward. The Mercury astronauts were continuing training, now with the Redstone. And NASA had announced that an astronaut (they still hadn't said *which* astronaut) would be launched on a Redstone in a suborbital flight, sometime that year. By early spring there seemed to be a fair chance that an American, as pilot or passenger, would be the first man in space.

Then, on April 12, 1961, came the news from Moscow. Major Yuri Gagarin, the son of a farmer, had orbited earth in a spacecraft called *Vostok 1*. A Russian in outer space. A human moon who was . . . *the son of a farmer*. If Sputnik had made Americans feel as if they'd slept through breakfast, then this was like being shaken awake from a ten-year coma. In the Soviet Union Gagarin became a national hero. He was interviewed by Khrushchev himself. Khrushchev asked him what he had been thinking about, out there among the stars, and the cosmonaut said, "I was thinking about our party and our homeland." Gagarin's reentry was the first time anyone had fallen so far so fast—a human being in a plummeting metal sphere inside a ball of fire, the white-knuckle ride to end all white-knuckle rides. The cosmonaut told Khrushchev that during reentry he sang the song, "The Motherland Hears, the Motherland Knows." Regarding capitalism and America's purpose there were now the faintest whisperings of doubt. Some of this doubt was registered in a poll conducted by the AP, which had its reporters call every Joe Smith in the telephone directories of ten cities, asking for a reaction to Gagarin's orbit. Some Joe Smiths said so what? Some said we'll catch up. A few said maybe the Russians have the right idea.

As it happened, Gagarin's flight came just as the House was discussing NASA's 1962 budget. Certain members of Congress were more upset than others. Representative James Fulton from Pennsylvania asked Hugh Dryden if the Russians had plans to explode a payload of red dust across the surface of the moon. When Dryden responded that intelligence sources had suggested the Russians might do exactly that, Fulton proposed that the U.S. consider responding by launching a rocket which would explode a payload of blue dust, complimenting the Soviet red and the natural white in such a way as to re-create the American palette in the celestial sphere. Fulton was exasperated. He said to administrator Webb and NASA in general, "Tell us how much money you need and we on this committee will authorize all you need." Some members of Congress wanted an all-out crash program to land a man on the moon, and NASA found itself in the strange position of having to turn down money it could not yet use.

In the White House, President Kennedy had been expecting the Russians to launch a man first. In fact, during his campaign he had said, "If a man orbits earth this year his name will be Ivan." Now he was simply resigned. He told the press, "We are behind . . . the news will be worse before it is better, and it will be some time before we catch up."

But if America was behind, it soon enough left the starting gate. On May 5, Navy Lieutenant Commander Alan B. Shepard was launched on a Redstone missile to an altitude of 115 miles. Fifteen minutes after lift-off, he splashed down in the waters off Bermuda. When he stepped out of the helicopter on the carrier deck, the medics, no doubt recalling the doomsday prognostications of psychiatrist D. G. Starkey, leaned forward with their arms ready to catch him should he fall. The man had been separated from the earth itself. Were his eyeballs rolled back? Was he twitching? Shepard just walked right past them. But he wasn't just walking, he was *strutting*. The man was absolutely exuberant. The first thing he said was "*What a ride!*" He strutted over to the capsule, reached inside, got out his helmet. It was like watching a healthy young man who had just thrown the shot that won the league championship for his college basketball team. Gagarin was reading from Khrushchev's script, but Shepard was laughing and making jokes! Suddenly Americans felt a little better.

Still, there was a good distance to go. Shepard's flight was suborbital. Only a lob out over the Atlantic launched from Cape Canaveral in an arcing pass 115 miles high and 302 miles long. To those in the know, it was only von Braun's Army Adam in a shiny new wrapper. Even the average American knew better than to confuse it with an actual orbit. A real orbit—and anything more ambitious than that—was a ways off. Some engineers were still considering there might be nearer, more easily accomplished space feats. One such possibility had occurred to men named Lovelace and Flickinger, and a woman named Jerrie Cobb.

Lovelace had arranged a third phase of tests for Jerrie at the U.S. Naval School of Aviation Medicine in Pensacola, and she was asked to report for them on May 15. Like the tests in Albuquerque, many of the Pensacola tests were designed to measure endurance, stamina, reflexes and dexterity. There had been no corresponding phase for the testing of the Mercury astronauts, in part because most had undergone similar tests in pilot school, and in part because some would be included in their actual training; at Pensacola the tests were often conducted under conditions meant to simulate actual space flight.

Jerrie was introduced to Lieutenant R. A. Carleton of the medical corps. For the next two days he oversaw a thorough check of her physical condition—electrocardiograms, electroencephalograms, X rays, and exercise tests. The third day Jerrie was fitted with a naval aviator pressure suit and helmet, and entered an airtight chamber which was depressurized to a simulated altitude of 60,000 feet. The physical fitness tests were more challenging. Jerrie stood five feet seven inches, and she was expected to scale a six-foot wall. She fell back the first time, but then managed a handhold, and was over. The sit-ups were easy enough, but the chin-ups were difficult, and nowhere was it more obvious that the tests had been designed for a male's muscle-mass. Carleton assured her that the tests would be modified for the women who might be tested later.

The airborne EEG was a system designed to measure brain waves while the subject was undergoing high g-loads in an aircraft. Because Jerrie was a civilian, the local authorities had to secure permission for her test ride from the Pentagon, and a wire was dispatched explaining that this was part of a program to test the differences between male and female astronauts. An admiral at the Pentagon, appreciating the occasional absurdities of military funding, wired back: "If you don't know the difference already, we refuse to put money into the project." But permission was granted, and Jerrie was fitted with a parachute and harnessed into the backseat of a Douglas Skyraider, where physicians fitted her head with a collar of eighteen small inward-pointing needles to measure her brain waves. Thus equipped, she spent an afternoon undergoing dives and rolls and loops. Afterward, she was invited to view the film from a camera mounted in the cockpit, and she saw her eyeballs bulge and then recede as the g-forces built up and then relaxed. Carleton said the response was perfectly normal.

Jerrie was reoriented in water survival techniques, life rafts, and helicopter harnesses. She was introduced to an aircraft crash simulator the locals called the "Dilbert Dunker." Designed to represent an open aircraft cockpit, it operated a bit like a roller coaster car; it would be propelled at considerable speed down an inclined track into a pool of water sixteen feet deep. Jerrie donned full flight gear, a life vest, and a parachute pack and climbed into the capsule. Then technicians strapped her into a nonoperative ejection seat. She inhaled and exhaled, deep steady breaths. She closed her eyes. Then she felt a violent jolt as the capsule was fired and, almost immediately, another shock as the nose crashed through the surface of the water. Water gushed and filled the capsule interior, rising above her knees and then her head. The capsule was on a kind of gimbal rig, and at the bottom of the pool it was inverted. Jerrie was holding her breath underwater, wearing a lot of flight equipment in a simulated cockpit sixteen feet below the surface. There were streams of bubbles all around her, but she could see the shape of the opening through which she'd make her exit. She knew she could not panic (they would be waiting for that), and knew that a person with healthy lungs could hold her breath for at least a minute. She told herself to move slowly, unhooked the harness and her parachute, and, careful to avoid hanging up on any edges, moved out of the capsule and let herself float to the surface, where emergency divers had been standing by.

The airborne survival check was designed to simulate a high-speed ejection from an aircraft. No such ejection was possible with the Mercury spacecraft, but its escape system would subject its passenger to similar g-forces. Bolted to the ground near the airfield was a mock-up of an aircraft cockpit, with an ejection seat mounted so that it would slide up and down a vertical steel track. Jerrie sat in the seat, and then a .50-caliber shell was exploded beneath her.

Immediately the seat was thrust upward with tremendous force, struck the stop and slowly slid back down the track. Afterward, a group of grinning naval officers handed her a card that named her a member of the "ejection seat club."

The final test took place in the "Slow Rotating Room." It was just that: a hollow cylinder eight feet high and fifteen feet in diameter that spun at variable speeds atop a forty-two-ton steel plate. Inside, it was furnished like a studio apartment, with a bed, chairs, tables, even a kitchen area. The purpose of the SRR was to induce disorientation; to a subject who did not adapt well, those comfortable surroundings might seem like the cruelest of jokes. Because the room had no windows there were no visual clues of motion, so although the subject's surroundings were spinning at a speed of 10 rpm, her brain continued to direct the body's movements as if the room were stationary. For half an hour, Jerrie was asked to perform exercises that would induce motion sickness—watching blinking lights, walking, throwing darts at a board and tennis balls into a basket, operating a series of knobs. She did well, and when she emerged from the room, she learned that she'd adjusted faster than had a group of incoming flight students.

By late May 1961, Jerrie had passed three phases of tests. If even half of the women she and Dr. Lovelace had found could make it as far as she had, and if the testing had the backing of both NASA's medical men and the Navy, then surely there was reason to hope that the next step would be a NASA-sponsored training program for women astronauts.

But in May 1961 NASA had other concerns. Kennedy's science adviser Jerome Weisner, no friend of the Mercury Program, had been suggesting that the U.S. should concentrate on unmanned scientific probes. They were less expensive, and in his opinion the scientific return was greater. Recently, Kennedy had assigned Vice President Johnson the responsibility of choosing a new NASA administrator. Kennedy had also made the vice president chairman of the Space Council, a group that would advise and assist the president with respect to policies and performance of all American aerospace activities, including those of NASA. Lyndon Johnson had more than a passing interest in the matter. As a senator, he had been present when Airman First Class Donald G. Farrell ended his week in a space capsule simulator (a sort of practice presidential reception for a practice spaceflight); and, as Chairman of the Senate Preparedness Committee, he had advocated the establishment of an independent space agency in 1958. Now Johnson was insisting that national prestige was at stake. Kennedy himself was undecided. So he ordered Johnson, Webb, and Secretary of Defense Robert McNamara to reevaluate America's efforts in space, and to make recommendations for NASA's future.

More than three hundred years earlier, aboard the Pilgrim ship *Arabella*, John Winthrop had spoken of the society they would build upon their arrival in the New World, saying, "We shall be as a City upon a Hill, and the eyes of the people are upon us." In 1961 Lyndon Johnson was beginning to realize that we were building another city upon a hill, and again the eyes of the people were upon us. The person sitting at the top of that rocket was a representative, a symbol. And that person—who he was and how he behaved and what he said—was going to have an effect back on Earth, in the streets of Bedford-Stuyvesant and Watts, in the halls of Congress, and in little crossroads towns in southern Mississippi.

Shepard's flight was only a fifteen-minute suborbital lob, but for Kennedy it meant that NASA was capable of a success. On May 25, Kennedy presented a special message to Congress on "urgent national needs." He set forth an accelerated space program based upon long-range national goals: early development of the Rover nuclear rocket, communication satellites, and a satellite system for worldwide weather observation. For all this he requested an additional $549 million for NASA over the new administration March budget requests. He saved the best for last: "I believe this nation should commit itself to achieving the goal, before this decade is out, of landing a man on the moon and returning him safely to the earth. No single space project in this period will be more impressive to mankind, or more important for the long-range exploration of space; and none will be so difficult or expensive to accomplish." If anything, Kennedy was underestimating the difficulty. To anyone who knew anything about the miserable capabilities of U.S launch vehicles in 1961, a manned lunar landing in eight years' time was more than a little optimistic. So far as anybody could tell, no technological breakthroughs were necessary, but they would need organization, coordination, and an assemblage of human and material resources on a scale that was unprecedented. There would be tremendous problems in propulsion, materials, guidance systems, structures, and life support. It would be more difficult and more expensive than the Panama Canal and the Manhattan Project combined. Worse, in 1961 a great deal about the moon itself was simply unknown. Some scientists were claiming that any spacecraft trying to land on the lunar plains would sink into an ocean of dust two miles deep.

Nonetheless, it was an inspiring speech, and few knew that long passages had been taken directly from a memo composed by Webb, McNamara, and Johnson. In early May, Webb and McNamara had submitted a draft to the vice president. It had concluded, "Now it is time to take longer strides—time for a great new American enterprise—time for this nation to take a clearly leading role in space achievement." Johnson read it with pen in hand, and in the space beneath that line he wrote, "which in many ways may hold the key to our future on earth."

23

1961

In May 1961, our future on earth was exactly what a lot of Americans were thinking about. The day before Shepard's flight, seven black and six white members of the Congress of Racial Equality began an expedition through the American South. Their intention was to challenge segregation at interstate bus terminals, at restaurants, at drinking fountains, and in public rest rooms. During those weeks in May there was violence and threats of greater violence. It was almost a miracle that no one was killed. But the freedom riders' plan succeeded. By early fall of 1961, the Interstate Commerce Commission would issue regulations requiring desegregation in all interstate terminals.

And so Vice President Johnson had good reason to think that minorities ought to be represented on the high frontier. He was not alone. Edward R. Murrow had similar ideas. Murrow was the correspondent whose short-wave reports from London under the blitz had mesmerized listeners across America. Late in his career, he had accepted Kennedy's appointment as director of the U.S. Information Agency. And on September 21, the day before the ICC issuance on desegregation, Murrow wrote a letter to NASA administrator James Webb. It was two sentences long: "Why don't we put the first non-white man in space? If your boys were to enroll and train a qualified Negro and then fly him in whatever vehicle is available, we could retell our space effort to the whole non-white world, which is most of it." Webb took Murrow's point, but politely responded that they had plenty of qualified candidates. Murrow didn't know that Webb had already received requests that a woman be allowed into the astronaut corps.

By this time Khrushchev and Kennedy, like Eisenhower, had both made statements asserting that human ventures into space ought to be peaceful and purely scientific. Similar sentiments were expressed in editorials and on university campuses, and some came from surprising sources. Two years earlier, James S. McDonnell Jr., founder of the firm that would build the Mercury capsule, had given an address entitled "The Conquest of Space: A Creative Substitute for War." By 1961 even the Atomic Energy Commission was forming a division to explore the possibilities of the "constructive" use of nuclear weapons as excavation and mining tools—it was called "Plowshare." And in late May of that year a group of scientists convened in Tulsa, Oklahoma, for the First National Conference on the Peaceful Uses of Space.

There was a banquet one evening, and at the head table was NASA's administrator James Webb. Webb had been a pilot in the Marines, earned a law degree from George Washington University, and had risen to prominent positions in the Truman and Eisenhower administrations. He had been NASA's administrator for just four months, and they had not been easy. During his second day on the job, the Russians announced that they had launched a space probe to the planet Venus. He had been under fire from many in Congress for many reasons—NASA cost too much, NASA was dragging behind the Russians, and so on. In partial response to these charges, Kennedy and his science adviser Jerome Weisner had appointed a panel to assess the Mercury Program. Its conclusions were mostly favorable. Still, for Webb, it had been a difficult spring.

That night at the conference banquet he was pleased to have before him an easier audience, and he was pleased to introduce the woman sitting near him. She had been mentioned in a paper Randy Lovelace had delivered that afternoon entitled "Future Problems of Man in Space," and she had just arrived from the Naval Air

Station in Pensacola, where she had completed the tests Lovelace called Phase III. Tonight Webb spoke of her success in that testing, and announced that he was appointing her as a consultant to NASA. She would be assigned to the Office of the Administrator as a consultant in manned space flight.

Jerrie hadn't been expecting the appointment, but she took it seriously. A few days later she traveled to Washington for a swearing-in ceremony at NASA headquarters. In attendance were Webb, director of Space Flight Development Abe Silverstein, and the director of the Office of Life Sciences, Colonel (USAF) Charles Roadman. Jerrie never thought herself particularly patriotic, but as she put her hand on the Bible and took the oath of office she felt the muscles in her throat tighten a little.

Upon returning to Oklahoma City, she took her first official action. In the absence of any specific directive from NASA, she composed her own mission statement. She would gather medical data relevant to sending women into space; she would educate young people about the value of space exploration and their contributions to it; and she would work toward making the first woman in space an American. Lovelace had told her that two years earlier the Soviets were researching the possibility of using women in space, so her recommendation to NASA was straightforward: Because the Mercury program would continue with six more suborbital ballistic shots like Shepard's, a seat on at least one of those rides should be reserved for an American woman.

The subject was generating concern at NASA's highest levels. George Low, now director of Spacecraft and Flight Missions, received a memo one day from G. Dale Smith. Smith had served on Lovelace's advisory committee and was now working under Roadman in the Office of Life Science Programs. He needed advice.

> There are a number of letters being sent to Headquarters by women who wish to become astronauts. I hope you will have ideas on answering these letters. The medical portion of a selection and training program is but a guide in selecting physically and mentally qualified people, and as you know, women fall within the physically qualified; therefore, there must be other valid reasons why or why not we are to use women in our space flight program.

Low had exactly one reason not to use women—none were jet test pilots. Despite the fact that the capsule's primary control was automatic, he still thought it a valid reason. And his response to Smith was exactly that.

Meanwhile, Jerrie herself was writing to various NASA officials, among them Webb, Silverstein, and Roadman, carefully outlining her rationale for a training program for women. Then she waited. In midsummer she opened a letter from Roadman.

> As you well know, the ultimate decisions in support of your views will have to be made by other than myself. I do think, however, that we should do some research in your areas of interest looking toward future requirements. At least, may I assure you, that I always endeavor to keep an open mind.

It was no invitation to training, but it wasn't an outright dismissal either. And if more research was needed first, it would come soon enough. Lovelace had suggested that the twelve women leapfrog Phase II, and move directly to the tests at Pensacola. The Navy could do it in July. Jerrie agreed. But Lovelace wanted to put a few more of the women through medical tests first, and he found that many had professional and personal commitments in July. He asked the Navy to postpone the tests, and they did—to the last two weeks of September. Now Jerrie was afraid that they were losing momentum. If they couldn't get to Pensacola yet, there was something she could do in the meantime. She asked Dr. Shurley if he would schedule women for isolation runs and associated exams at the Veterans Administration Hospital in Oklahoma City, the same tests she had undergone in September of the previous year. He would. Of the ten women who had passed the tests in Albuquerque, two—a charter pilot named Rhea Allison and a flight instructor named Wally Funk—made themselves available for the psychological stress tests.

Jerrie invited Rhea and Wally to stay with her while they underwent the tests. She was renting a three-story house in one of Oklahoma City's older neighborhoods, and she fixed up an upstairs bedroom. She found bedspreads adorned with rocket ships, and papered the ceiling with celestial wallpaper. Her optimism was justified by many things. The letters she had received after her tests were made public and her reception at the conference in Tulsa suggested that Americans would support a plan to launch a woman into space. Lovelace was trying to push his testing program to another stage. The Navy was interested enough to run tests on the women, and NASA's Lewis Research Center had allowed Jerrie to ride the MASTIF. Perhaps most significantly, Jim Webb and other administrators at NASA seemed reasonably receptive.

Jerrie met Rhea Allison at Wiley Post Airport near her office. Rhea had flown up from Houston in her employer's Piper Comanche. She had 1,500 hours, and held a commercial license, an instructor license, and ratings in multiengine, single-engine, and seaplane. She had towed gliders for the Air Force Academy at Black Forest Gliderport, and passed the tests at Albuquerque in the spring. Rhea spoke rarely, but she had a perpetual smile that bespoke comfort with her place in the world. When people commented on her reserve, she would say that we were given two ears and one mouth so we would listen twice as much as we spoke.

On Wednesday Jerrie met Wally Funk. Wally was young, only twenty-two, a test pilot for a California Flying Service. She was rated as a flight instructor, and she had 3,000 flying hours. She had passed the tests at Albuquerque a few months earlier. Her given name had been Mary, but she never used it. In 1961 most of Wally's experience as a pilot was still ahead of her, but it was hard to forget a woman with a name like Wally Funk, and few did.

The tests began again.

Interviews, Rorschach (ink blot), thematic perception (the stories suggested by pictures), draw-a-person, sentence completion, self-inventory based on 566-item questionnaire, officer effectiveness inventory . . .

. . . personal-preference schedule based on 255 pairs of self-descriptive statements, preference evaluation based on 52 statements, determination of authoritarian attitudes, peer ratings, and determination of the question Who am I?

. . . the Wechsler Adult Scale, Miller Analogies, Raven Progressive Matrices, Doppelt Mathematical Reasoning scale, engineering analogies, mechanical comprehension, Air Force Qualification Test, Aviation qualification test (USN), space memory, spatial ori-

entation, Gottschaldt Hidden Figures, and Guilford-Zimmerman Spatial Visualization . . .

Meanwhile, the last two women to undergo Phase I in Albuquerque had written directly to Dr. Lovelace, volunteering themselves. Gene Nora Stumbough, who had heard about the program from Wally Funk, was a twenty-four-year-old student making her way through the University of Oklahoma by giving flight instruction. She had 1,400 flying hours, multiengine, and ground instructor ratings. Gene Nora had never believed the tests would evolve into a training program. She was told they were merely involved in an important research program to set up parameters for future women astronauts. But for Gene Nora it was enough to participate in astronaut tests. She was a bit surprised, and at first a little intimidated by her roommate, Jane Hart. Jane, who had received word of the tests from Bernice Steadman, was thirty-eight, a mother of eight, the wife of much-respected Senator Phil Hart, a captain in the Civil Air Patrol, and a Detroit millionaire with her own six-passenger Aero Commander. Jane had dark, short-cropped hair, and features some might call severe. She would joke that her eight children, four girls and four boys, were a perfectly democratic and equal representation of sexes.

. . . Density of the body was determined by weighing the nude body in water after maximal inspiration followed by exhalation of a measured amount of air . . .

By July 1961, twenty-six women had been tested, and thirteen were judged physiologically suited to spaceflight. Three of the thirteen, who had passed Phase II, were judged psychologically fit as well.

In the summer of 1961 Jerri Sloan was back in Dallas, managing Air Services and caring for Candy and David. When she could, she did other work, after hours. That summer she helped arrange a convention to honor the Business Pilots' Association, whose members had freely given their services as private pilots to local humanitarian efforts—flying medicine around the state, flying victims of tornadoes to shelters, and so on. Jerri was a little surprised to learn that someone had invited Jackie Cochran as a speaker. Jerri recognized Cochran's considerable accomplishments, and knew they weren't all behind her. Still, as far as Jerri was concerned, Cochran's records were made possible not so much by her aviation skill as by the fact that her husband owned the companies that built the planes. In Jerri's eyes, Cochran was a self-promoter and grandstander, and had done more for herself than she'd ever done for the cause of women in aviation. That particular night, Cochran took the podium and began to talk about aviation during the war. She said, "Hap Arnold told me his men were afraid to fly the B-26, and I told Hap, then my girls will do it." To hear her tell it, she'd practically won the war herself. Jackie may not have realized that the business pilots in the audience had not always been business pilots. Many, in fact, had flown B-17s and B-24s with the Eighth Air Force and the Fifteenth Air Force. Some had been wounded badly, and several had seen their fellows die before their eyes. Tonight they had been invited to be honored for charity work. But Jackie just kept talking . . .

Jerri Sloan was sitting on the dais near the front of the hall, and the more Jackie talked the more Jerri wanted to sink beneath the table and then through the floor. It seemed to her that Cochran had insulted and angered just about every person in that room.

Jerri had vowed that she would never associate with Cochran again, but there was no point getting into that now. The speech was over and the audience had left; it was too late to do anything. Jackie approached Jerri, and began to talk to her about the testing in Albuquerque. In April, Cochran herself had tried to pass the tests, but her circulatory and intestinal problems were severe, and Lovelace told her what she must have known, that she could not hope to pass even Phase I. Cochran had helped Lovelace choose candidates, and had paid the expenses of the women who could not afford their own. Beyond this, the precise nature of Cochran's relation to the women's tests was clear to no one. But it seemed to Jerri that she expected to manage any training program that might ensue. Cochran said something to Jerri about the upcoming tests for the women at Pensacola. Because Jerri hadn't received any news about Phase III, she said, "Well, Jerrie hasn't notified us of—" Before she could finish Cochran let fly a string of epithets that ended, "Jerrie's not running this program—*I* am." For a moment there was quiet. Then Jerri looked up at Cochran and said, slowly and evenly, "Well that's news to me."

By late summer Jerri Sloan and the other women were given formal instructions from Randy Lovelace.

> August 21, 1961
>
> Enclosed is our check for $200 to cover your expenses to and from Pensacola and your maintenance while there as provided by Miss Cochran. Between now and September 17, I suggest that you study the FAA manual, mathematics, theory of flight, meteorology, and things that have to do with design of aircraft and engines. There will be some examinations in all of these subjects.
>
> W. Randolph Lovelace
>
> enclosure
> cc: Jackie Cochran

Jerri, for one, did not appreciate that "cc:" But whether Jackie Cochran had time to read her mail in the late summer of 1961 is an open question. On the day after Lovelace wrote the letters, she was preparing for a record attempt, flying a Northrop T-38 at an altitude of 50,000 feet and an average speed of 863 miles per hour.

Since breaking the sound barrier in 1953, Jackie had set some aviation records for women, but had had few opportunities to set absolute records. One reason was that she relied upon Yeager's influence to gain access to the jet aircraft she needed, and preferred him as an instructor. For many of those years Yeager was stationed elsewhere. But in 1961 Yeager was again at Edwards, recently appointed to head the new Aerospace Research Pilots School to train military astronauts. He was again ready to vouch for Jackie's skills to whomever asked, and again ready to act as instructor. The aircraft in question would be a Northrop T-38, a sleek new Air Force jet trainer, powered by twin General Electric turbojets, each providing four thousand pounds of thrust. In the summer of 1961, as in 1953, there were some bureaucratic and legal hurdles. Although there were several T-38s at Edwards, they were the property of the Air Force, and the Air Force would not allow civilians to fly its planes. So the Air Force agreed to lease one T-38 back to Northrop. Northrop would allow only its own pilots to fly its planes. But the aircraft firm was planning to develop the T-38 into a fighter version it would call the F-5, and knew that records would assist sales, perhaps especially records set by Jackie Cochran. So in the summer of 1961 Northrop Corporation was persuaded to hire Jackie Cochran as a pilot. Again she moved into accommodations on Edwards Air Force Base, was fitted with an oxygen mask, and made test runs in an altitude chamber. Then she and Yeager and General Ascani waited for clearance from the Pentagon.

At Edwards it was an exciting time. The Aerospace Research Pilot School was designing its courses and selecting applicants, and the X-15 had been flown in various configurations for more than a year. By August 1961 there had been thirty-eight X-15 flights, one every few weeks.

On August 4, Jackie had clearance. She began practicing touch-and-go landings in a T-38, with Yeager in the backseat. A few days later she flew it alone, while Yeager flew alongside in an F-100. She accelerated to Mach 1.2, tried stalls, shut down one engine, and restarted it in flight. Although she made a few mistakes, she was growing comfortable with the feel of the plane. On August 10, she climbed to 40,000 feet and for a few minutes chased an X-15, drop-launched at the same moment. Almost every day from August to October, she flew that T-38 in preparation for various records. She made practice runs on straight courses from Edwards, three hundred miles across the desert and back. Sometimes she flew closed courses, imaginary circular tracks 30,000 feet in the air, with smoke curls for pylons. If she broke a record or if the day went well, that evening there would be steak dinners at the house she was renting, with Yeager and an X-15 pilot or two. Sometimes Floyd would come up from the ranch, and they would talk long into the night.

By early fall, Cochran would set eight major records in that T-38, including speed over a one thousand-kilometer closed course, and distance in a straight line. She would claim the world speed record for women at 842.6 miles per hour, and set a new woman's altitude record of 56,071 feet. After the latter flight, Northrop representatives presented her with a bouquet of yellow roses, and newspapers across the country picked up the press release. Although the release seemed to ask for embellishment or explanation, it was hard for the newspapers to know what else to say. The newspapers could not much discuss the nature of the aeronautical records she had broken, because few readers would understand or care. Neither could the newspapers do much to glamorize Jackie Cochran herself. Had she been twenty or thirty years younger, they might have made her the object of female envy and male desire. But in 1961 she did not look like the glamorous aviatrix of the 1930s, and she did not look like an ornament from one of Hollywood's recent adventures into outer space. In 1961 Jackie Cochran looked like a fifty-year-old woman. And most newspapers simply printed the release without comment.

Meanwhile, the public knew nothing at all of twelve other women pilots, who were at that moment quietly readying themselves for tests at Pensacola. The Navy would put them through it together; they would stay on base at the Bachelor Officers' Quarters. For all, it would mean rearranging a schedule, but for some it would be more difficult. Sarah Gorelick's work at AT&T involved rerouting certain customers who had lines in the same cable pairing. If telecommunications in certain parts of the country were destroyed by nuclear blasts, signals could take another route. AT&T's workers had spent years consolidating, and now, because of the Soviet threat, they were working overtime. But Sarah had already taken time off for those ten days in Albuquerque, and her supervisors were saying no more time off. She resigned. In Norman, Oklahoma, Gene Nora asked her supervisor at the university if she might have two weeks off from her flight instructor work for the tests. As it happened, those two weeks were the first two weeks of classes. She could leave for the tests, he said, but he'd have to replace her. She resigned too. Gene Nora and Sarah and Irene and the others studied the FAA manual and flight theory, when they had time. Jerrie had sent them liability waivers required by the

Navy. Each of them signed and returned them. Most would arrive in Pensacola on the afternoon of Sunday, September 17. Jerrie was anticipating this as the moment all the women would be together for the first time. It would be quite a sight—a meeting of thirteen American women who might become astronauts.

Two days before they were to leave Jerrie got a phone call from Randy Lovelace. "Jerrie," he said, "I just had a call from Pensacola. The testing is off." Jerrie was quiet for a moment, a little stunned. "What happened?" Lovelace's voice came back tired, defeated: "I'm afraid that's all I know."

So Jerrie called Pensacola. She talked to a great many administrators at the Naval Air Station, she sent telegrams, and she flew to Washington to meet with Silverstein or Webb or anyone who would talk to her. Finally, she found an answer—of sorts. In order to justify the expense of the tests, the Navy had needed a formal expression of interest from NASA called a "requirement." A few weeks earlier, Vice Admiral R. B. Pirie, deputy chief of Naval Operations, had asked NASA if the agency had requirements for the testing of women. His request had provoked a few days of discussions between the staff of the Navy and NASA, the upshot of which was that despite the manifest interest of Jim Webb, NASA would not pay for the tests, and was in no mood to issue the requirement.

And so before they were scheduled to leave, Jerri Sloan and Sarah Gorelick and Gene Nora Stumbough and nine other women received telegrams from Jerrie they simply did not understand:

> REGRET TO ADVISE THAT ARRANGEMENTS AT PENSACOLA HAVE BEEN CANCELED. PROBABLY WILL NOT BE POSSIBLE TO CARRY OUT THIS PART OF THE PROGRAM. YOU MAY RETURN ADVANCE TO LOVELACE CARE OF ME. WILL ADVISE OF FURTHER DEVELOPMENTS.

But it was clear that Jerrie was trying, because a few days later there was a note from her secretary: "Miss Cobb has just informed me from Washington that she has been unable to reverse decision regarding Florida testing again." In early October NASA deputy administrator Hugh Dryden wrote to Admiral Pirie, confirming NASA's position: "NASA does not at this time have a requirement for such a program."

But from a higher level Jerrie was still getting a more equivocal response. In late September she opened a letter from Jim Webb. It was a kind of apology—"I know we have been slow in working out arrangements for utilizing your services. Brainerd Holmes is coming in as director of Manned Space Flight programs on November 1, and my proposal would have to be deferred until then." For Jerrie Cobb it was a confusing autumn.

But for the nation as a whole, international events had been conspiring to make some things terrifyingly clear. On July 21, Astronaut Gus Grissom made a second suborbital flight, identical to Shepard's, a fifteen-minute arcing pass one hundred miles high and three hundred miles long. The capsule sank shortly after splashdown, but Grissom himself got out in time. NASA was pleased that the Mercury Program was progressing as planned. Then in early August, came another newswire from Moscow. The Russians had done it again. *Vostok 2*, carrying Major Gherman S. Titov, had circled the earth seventeen times, a whole day in space. The Russians had taken another giant step. Soon it became clear that Khrushchev had timed the flight precisely to distract world attention from activities which had very little to do with what he called "the service of peace and scientific progress." On August 12, a mere five days after Titov's return, Red Army engineers and German border guards moved prefabricated concrete barricades and barbed wire into place along the border between East and West Berlin.

On the first of September the White House announced that the Soviet Union had resumed testing of nuclear weapons. It was the first known nuclear test by anyone since the fall of 1958. In the next two weeks there were five million requests for the Department of Defense's thirty-two-page pamphlet, *The Family Fallout Shelter*. The Kingston, New York, common council called for bids to equip fifteen acres of limestone cave as fallout shelters. Now there was an edge of sadness in people's voices. One man, watching a truck carrying a large tar-coated concrete cylinder with vents, said, "A year ago I'd have snickered."

For weeks it was like a great chess game, move and countermove. On October 5, an Atlas missile carrying a dummy warhead was fired nine thousand miles across the Atlantic Missile Range. It splashed down in the Indian Ocean. On October 27, there was a sixteen-hour confrontation between Soviet and U.S. tanks. And on October 30, the Soviet Union exploded a fifty-five-megaton nuclear device. In early December Kennedy met with members of the National Security Council, and said that he would have to resume testing.

At the end of that year, there was a rather different kind of interchange between two representatives of these nations. In fact, one American woman received an offer of sorts to fly into space, from an unlikely source. It occurred at the Kremlin's New Year celebration, and for the Russians there was much to celebrate. Yuri Gagarin, the first man in space, was a guest of honor. He was introduced to the U.S. ambassador to the Soviet Union, Llewellyn Thomson, and his wife, Jane. As it happened Jane Thomson was uncommonly attractive, and the cosmonaut's politics were no hindrance to his appreciation for the fairer sex. This time Gagarin was not reading from Khrushchev's script, but rather behaving as any healthy jet test pilot might behave in an off-duty moment. He moved a little closer to Mrs. Thomson, and asked, "How would you perhaps like to go into orbit with me?" Mrs. Thomson smiled, and gave an answer which was both utterly truthful and decidedly diplomatic. "I'd be frightened to death," she said. "Besides, I'm not in training."

In 1961 it was one of the kinder communications between the earth's two spacefaring nations. At that very moment, it was late afternoon in Washington, D.C. Most of its citizens were leaving work early, thinking about dinner with friends or an office New Year's Eve party. But a few people in that city would work through the night and on into the morning. They were in a room thirty feet beneath the Pentagon, in constant contact with military bases in North Dakota and Utah. And at those bases were 294 Atlas missiles, and 630 long-range bombers—B-52s and B-47s, a third of which were airborne at any given moment. As the earth turned, the steppes around Moscow moved toward dawn, and the east coast of North America moved into darkness. By a coincidence of the natural forces that create geography and the political forces that determine national boundaries and time zones, the Soviet Union was a nation always half a day ahead. As the year drew to its close, the symbolism seemed altogether too perfect.

24

"THE GIRL," REDUX

It was 1962, already six years into the Space Age, and the Russians still had every space first. The first artificial satellite. The first live animal into orbit. The first man in space. The first man in orbit. The second man in orbit. They had even put an unmanned probe out past the moon, and sent back a blurred photograph of its far side. It was true that the United States had launched more satellites and probes—whole families of Vanguards, Discoverers, Explorers, Pioneers, Samoses, and Tiroses—and they were more sophisticated than the Russian Sputniks. Nonetheless, much of the public was uninterested in anything that didn't have a human being inside. And the public was not alone. Hanson W. Baldwin, the influential military affairs correspondent for *The New York Times*, had cautioned against an emphasis on science at the expense of political realities: "It is not good enough to say that we have counted more free electrons in the ionosphere than the Russians have. . . .We must achieve the obvious and the spectacular, as well as the erudite and the obscure." For at least a year there had been rumors that another kind of Russian spectacular was imminent, and now there was new support for those rumors. On the radio channels the Russians usually reserved for cosmonaut training, U.S. intelligence had heard women's voices.

But an *American* woman in space was still an utter fantasy. And on some days even an American man in orbit seemed unlikely, at least in the near-term. The rockets themselves weren't always getting there. The Atlas missile, with which NASA was planning to orbit a man, was still iffy. One had exploded on July 29, 1960, and on April 25, 1961, another was destroyed forty seconds after takeoff by the Range Safety Officer because of a guidance malfunction. Only a few weeks earlier, on December 15, an Atlas-Able exploded seventy seconds after launch. And in the ready room of Air Services in Dallas, Jerri Sloan heard a joke. "How does a kid from Cape Canaveral learn to count?" "4-3-2-1-*damn*."

When the letter about Pensacola had come to Jerri's home in Dallas, she had said "To hell with it." The women went back to jobs, husbands, and children. Sarah Gorelick found work as an accountant back in Kansas City. Gene Nora Stumbough had college debt and no immediate prospects, but she managed, and soon found work as a flight instructor at Oklahoma State. Irene Leverton took a course for air transport ratings in Long Beach, California.

Jerrie Cobb, meanwhile, was still doing what she could to convince NASA to let a woman go into space. She was acting as a semi-official liaison between the women who had passed the tests at Lovelace and NASA. On a regular basis she wrote to the women about the higher-ups she was trying to meet, and who seemed amenable. She began those letters with "Dear Fellow Lady Astronaut Trainees." In truth, it was a rather inaccurate description: They had not been invited to enter actual astronaut training, and so they were "trainees" in no sense NASA would appreciate.

But life went on, and most days Jerrie would try to put the astronaut business out of her mind. There was still work at Aero Commander, and she was flying demonstrations around the country. Then Shepard or Grissom or one of the astronauts who had yet to fly appeared on television, and she stopped to listen. She felt great envy. Of course Jerrie and the other women wished the astronauts all the best, but sometimes it was hard. Some of the astronauts had made comments in public that dismissed the idea of women in space. And there was a remark about how women might be on board because NASA made a small weight allowance for recreational equipment. There were a lot of laughs over that one. But when Jerri Sloan heard it she did a slow burn: "*I am nobody's recreational equipment . . .*"*

In September 1961 *McCall's* magazine published an account of the Lovelace tests authored by Marion Deitrich. Marion had written the piece over the summer of 1961, as she and the others were waiting to hear from the Navy. It was an interesting article, and an unusual one, bordered as it was by ads about refrigerators and seamless stockings. It was the first time the story had appeared in the national press since the AP wire in January. Soon enough, local papers followed suit. Suddenly, there was attention, not all of it polite. In fact, there was a good deal of outright hostility. Over the next few weeks Jerri Sloan heard and overheard a number of impolite remarks from male pilots.

They had lived for thirty years in a world in which a woman might have to work twice as hard as a man for the same recognition. As pilots, they not only had to fly better than male pilots, they also had to be ladylike about it. Irene Leverton had been fired from at least one position merely because she asked for equal pay. But all those experiences did not prepare them for the sheer ferocity of the anger unleashed upon them now. They were getting hate mail, some of it from women. Some were saying they should be taking care of their children. As far as Jerri Sloan was concerned, this was the unkindest of cuts. Nothing was more important to her than her children—and it occurred to her that nobody was questioning Gus Grissom about his paternal responsibilities because he was spending more time in the capsule simulator than with his wife and his two sons. The women called each other and said *Don't let*

*The women had heard the joke secondhand, attributed to various NASA officials, and even to astronauts. The surviving textual evidence is from a speech by Werner von Braun, delivered at Mississippi State College on November 19, 1962: "Another question that I am frequently asked is this: 'Do you ever plan to use women astronauts in your space program?' Well, all I can say is that the male astronauts are all for it. And as my friend Bob Gilruth says, we're reserving 110 pounds of payload for recreational equipment." Robert R. Gilruth was at the time director of the Manned Spacecraft Center. However Gilruth may have intended the remark, von Braun's aside was more a lapse of taste than a dismissal of women astronauts. The text continued, asserting that women and men of all backgrounds would eventually fly into space.

it get to you. Jerrie Cobb herself just smiled and said it wasn't important. She would talk about the next step, or who at NASA she was trying to meet. But for the others it was pretty hard to take. They'd see the astronauts on television and in *Life* magazine. The men could do no wrong, and the women were getting hate mail. It didn't make sense. It was like a blizzard on a warm summer afternoon. But the American public's feelings regarding these thirteen women was changeable. They received many letters of support, too. And sometimes the blizzard would stop. Because every now and then it would occur to somebody that these thirteen women, by all appearances entirely sane, wanted to ride an intercontinental ballistic missile in the place a nuclear warhead is supposed to be.

NASA's original plan for Project Mercury had been to put all the astronauts through a suborbital lob like that of Shepard and Grissom. Seven astronauts, seven Redstones. But after Grissom's flight it became clear that the Mercury spacecraft had been sufficiently tested in suborbital flight. And then came the twin *Vostoks*, and NASA began to feel that two suborbital lobs was enough. It was time to put an astronaut on an Atlas. It was time to launch an American into orbit.

On November 29, NASA completed a successful test flight of a Mercury/Atlas carrying a 39-pound chimpanzee named Enos. Enos rode two orbits around the earth. The spacecraft, booster, communication and tracking network, and recovery operation performed almost flawlessly. The system was ready. NASA announced that the first American in orbit would be Marine Lieutenant Colonel John Herschel Glenn. Glenn was a hero made to order. He had flown fifty-nine combat missions in the Marshall Islands and ninety in the Korean War. He had been awarded the Distinguished Flying Cross five times and an Air Medal with eighteen clusters. On top of all this, he was a small-town American. Someone once said that Glenn behaved each minute of the day as though every impressionable youth in America were watching.

Glenn's flight was postponed again and again. It was scheduled for January 16. There was a problem with the propellant tanks, and it was rescheduled for January 23. In the intervening week hundreds of reporters and thousands of people crowded the beaches and motels around Cape Canaveral, down in Cocoa Beach and up in Titusville and all along Route A1A. Some had trailers and parked them on the beach within sight of the rocket. But on the big day a layer of clouds shrouded southern Florida. The launch was postponed to the next day, and there were clouds that day and the next and the next.

Finally, it was February 20, a Tuesday morning, and when the newspapers and television and radio people heard the launch was on again, they had doubts. But as the morning brightened, the sky began to clear. In the area near the Cape, there were thousands of spectators, many in official capacities. In 1960 the Fédération Aéronautique Internationale had adopted rules to govern the award of official records for manned space flight, ruling that all such claims must be supported by eyewitnesses. Jerrie Cobb was present at Cape Canaveral as the FAI's official observer.

Across the United States that winter morning, everyone was listening or watching. Those who could not stay home from work brought portable radios and televisions with them. Finally it happened. Three tongues of bright flame shot from beneath the Atlas missile, and a thunder drowned out all other sound. Astronaut Scott Carpenter spoke to Glenn over the radio link, and the nation listened in. With the faintest quaver in his voice, Carpenter said, "Godspeed, John Glenn." *Godspeed?* The language of 1962 suddenly seemed woefully inadequate for this noble enterprise, and Carpenter had reached back into an older lexicon. *Godspeed.* It was as though America had become some luckless medieval village cursed by four years of failed crops, its last hope of redemption a knight sent eastward to find the Grail. And so as that slender white rocket rose into the bright morning sky on a pillar of flame, Americans said silent prayers and stood there before their televisions with tears in their eyes. And Carpenter's words resounded across the country like a benediction from the village's high priest. *Godspeed, John Glenn.*

Of course, not everyone had tears in their eyes. Jerri Sloan, watching the flickering image on the television in her home in Dallas, said aloud to no one in particular, "No hard luck, ol' buddy, but I hope you bust your ass."

25

GLENN

John Glenn did not bust his ass. That Atlas missile rose into the Florida sky and arced eastward. The crowds at the Cape and the Americans in their living rooms and in their classrooms watched the flame get smaller and smaller until they could no longer see it. They heard John Glenn report that the capsule he called *Friendship 7* separated from the Atlas missile and he was in orbit. A few hours later Glenn returned from outer space a national hero the likes of which the country had not seen since Lindbergh. After two days of debriefing there was a press conference, then a weekend in Key West with the family. Then Glenn, with the president, flew to Washington aboard Air Force One, where he addressed a joint session of Congress.

Because John Glenn's flight tended to overshadow all else that happened on February 20, 1962, few noticed that on the same day NASA put into effect its official policy on equal employment opportunities for women. In a memorandum to all NASA employees, James Webb declared, "It is my intention to take positive steps to ensure equal opportunity for employment and advancement for all qualified persons on the sole basis of merit and fitness without discrimination on the basis of sex. I expect NASA employees at all organizational levels to give full support and cooperation to this program." In the coming months, the agency would make a concerted effort to recruit female technicians, mathematicians, and engineers. But no plans were made to hire women as astronauts.

For nine months, since the tests in Pensacola had been canceled, Jerrie had done everything she could to persuade Jim Webb to reconsider his decision, and Webb had remained polite but firm in his refusal. He was not philosophically opposed to sending women into space; he simply insisted that now was not the time. But Jerrie feared that now might be the only time. She could not forget Lovelace's words three years earlier—*we had indications . . . the Russians are planning to launch a woman.* The Russians never announced flights or identified cosmonauts before the launch, but sometimes they dropped hints. As recently as December of the previous year, Lieutenant General Kamanin, an aerospace engineer making a world tour with Gagarin, had said that the Soviet Union was planning to launch a woman into space.

Jerrie had concentrated her lobbying efforts on NASA itself, and on the American public. But by the spring of 1962 Jane Hart had come to share Jerrie's impatience and her frustration. Jane Hart, wife of a senator, knew how to appeal to politicians. And now she did. She wrote letters to each of the twenty-seven members of the House Committee on Aeronautics and Science explaining the importance of a space "first" that the U.S. could claim, and the need for examples for American girls. All but one responded that they had read the material and found it of interest. By March there was real progress. Jerrie and Jane had appointments to meet with several members of Congress. They also had an appointment with Lyndon Johnson, chair of the Space Council and vice president of the United States.

If Kennedy was beginning to appreciate the political uses of astronauts, Lyndon Johnson was way ahead of him. Johnson looked at the image of John Glenn on the television and said, "If only he were a Negro." Kennedy used to tell the story to his friends, and that would be the punchline—Kennedy's point being, more than anything, that Johnson never gave the politics a rest. Here we are looking at a future in the stars and this uncouth Texas schoolteacher is thinking of voting precincts. But the Kennedy administration was sensitive to issues of race, and in early 1962 Attorney General Robert Kennedy asked General Curtis LeMay, now head of the Joint Chiefs of Staff and Chief of the Air Force, to ensure racial integration within the astronaut corps.

At the time, the Air Force had begun to recommend graduates of its Aerospace Research Pilots Course to NASA, who relied heavily upon those recommendations, and would accept thirty-eight of the graduates over the next six years. LeMay ordered that a black pilot be admitted to the course. In fact, there was one black applicant—Air Force Captain Edward J. Dwight Jr., a B-57 pilot with 2,000 hours of experience in high-performance jet aircraft. In August 1962 he was enrolled in Phase I of experimental test piloting at Edwards; he was graduated the following April, and among the fourteen admitted to the Aerospace Research Pilots Course. Dwight was by all accounts fully qualified for the program, but was regarded by the other pilots as the "Kennedy guy," the one who had been admitted through White House influence. In the spring of 1963 Dwight would be one of a hundred and thirty-six astronaut applicants nationwide, and among twenty-six recommended to NASA by the Air Force. By fall of that year, fourteen men would be selected for astronaut training as part of Group III. Dwight would not be among them.

But in early 1962 Johnson's thoughts on black astronauts were also more high-minded than the president's joke implied. To Lyndon Johnson, Glenn was the embodiment of the American dream, a living example of the adage that any American, no matter how humble his origins, could reach extraordinary heights of achievement. And in 1962 the country had great need of extraordinary achievers. Reports appeared which stated that pupils in Soviet grade schools were studying optics and quantum physics, and the Soviet Union was graduating twice as many engineers and scientists as the United States. After Sputnik, Senator William Fulbright of Arkansas claimed, "The real challenge we face involves the very roots of our society. It involves our educational system, the source of our knowledge and cultural values." In 1958 Congress had enacted the National Education Defense Act to provide federal aid for improved teaching in science, mathematics, and foreign languages. In December 1960 the National Science Foundation had announced grants totaling $22.7 million to support summer institutes for twenty thousand teachers of science, mathematics, and engineering in high schools and colleges.

By 1962 it had occurred to Jane Hart that there was perhaps a simple solution to this crisis, born out of the recognition that a large segment of the American population was vastly underutilized. Few had thought to encourage American girls to pursue careers in sciences and engineering, but Jane had been corresponding regularly with women college students, majoring in precisely those fields, and she couldn't help but note a tone of discouragement. The situation called for an example—an ordinary, representative American allowed to achieve something extraordinary. In the spring of 1962 Lyndon Johnson knew that America needed a black John Glenn. And Jane Hart knew it also needed a female one.

On March 15, Jerrie and Jane met with Lyndon Johnson for half an hour. They explained not only their reasons for wanting to be astronauts but their qualifications—they had passed the medical tests and they were experienced pilots. There was every indication that the women would perform as well as the men had; such a program would require no major adjustments of the manned spaceflight operations already in place, and so on. Johnson listened politely. Then he stood, led them to the door, and said he would certainly give the matter some thought. On April 23, he sent Jerrie Cobb a letter in which he expressed his pleasure in meeting with her and congratulated her on her "cogent arguments in favor of testing and training women as astronauts." But he concluded:

> The choice and training of the individuals who will make space flights is quite appropriately left to the operating agencies in the program. However, I see no reason why preparations should not be made for testing and training individuals who have the required physical and mental capabilities, regardless of sex.

And that was that. Johnson would support the cause wholeheartedly, but he would not advance it, and asked that they defer to

NASA's judgment. This Jerrie could not do. While in Washington, she had also tried to schedule a meeting with James Webb. Webb was unavailable, but he arranged for her to meet with Hiden Cox, NASA's assistant administrator for Public Affairs. To Jerrie, this was tantamount to saying that the matter of women's future in aerospace was not a policy issue but a public relations issue. Indeed, there was something to her suspicions. Unlike Glennan and Dryden and the first generation of NASA administrators, Webb was not an engineer. He was a Washington insider, a bureaucrat, the closest thing to a politician without actually being one. Indeed, now it was beginning to seem that his appointment of Jerrie had also been political.

Jerrie and Cox met in Webb's office. Cox heard her out politely and then said equally politely that there was nothing he could do for her. He had no authority over matters of policy. Shortly after their meeting, Jerrie opened a letter with a return address in NASA's Public Affairs Office. It was from Cox.

> I believe the upshot of our discussion in Mr. Webb's office of several days ago was that the future use of women astronauts is possible, but at what time in the future programs is another matter entirely. Despite the manifest interest in your proposal from audiences who hear you speak, I am afraid that at the present we cannot undertake an additional program training women to be astronauts.

She was disappointed, but she took heart in the support shown by friends, and the audiences of civic groups and women's groups. On March 30, Jerrie wrote again to Webb, and for the first time she allowed a hint of irritation to show: "If you do not think this proposal to be important enough to warrant your attention and it is out of Dr. Cox's authority, would you please tell me what you would like for me to do?" Near the end of the letter she softened her tone a little and concluded, "Please let me help." Meanwhile, Jane arranged meetings with congressmen George Miller and Victor Anfuso, both of whom served on the House Committee on Science and Astronautics. Like Cox, they listened politely and promised to give the matter some thought.

On May 24, Scott Carpenter became the fourth American in space and the second to orbit the earth. On the next day, in a conference hall at around 1:00 P.M., a business luncheon was winding to a close. On the podium in the front of the room a man stood and cleared his throat gently, and the guests at the luncheon settled into their chairs to listen to the keynote address, which on this day was being given by an attractive young woman from Oklahoma named Jerrie Cobb. Miss Cobb was to speak on America's future in outer space. With the completion of yet another successful American mission, this topic was of even more than usual interest. The fact that the acknowledged expert seated in their midst was a young attractive woman made it all the more thrilling. It seemed to mean that this venture into outer space really was every American's venture. In fact, that was precisely what Miss Cobb was here to talk about.

As the man at the podium said the usual things about being honored to have such an esteemed guest, Jerrie sat quietly beside him, battling a case of nerves, trying not to notice the hundred pairs of expectant eyes. She had done this so many times by now that it was often difficult to remember what city she was in, and what group she was addressing. She would never learn to like public speaking, and today she could take comfort only in the fact that she had given the speech so often that there was little chance she would stumble over it. She expected the audience would be receptive; they nearly always were. But Jerrie wanted her audience to believe that sending American women into space was a goal worth pursuing now. This meant that the speech had to be persuasive.

Jerrie rose. The audience applauded a little, and the man moved away from the podium and Jerrie stepped over to take his place, took a quick deep breath and began. She made a few remarks about all that the space program had already accomplished—how she admired the achievements of Shepard and Grissom and Glenn and Carpenter. Then she paused and said, "First, let us recognize that we women pilots who want to be part of the research and participation in space exploration are not trying to join a battle of the sexes. We seek only a place in our nation's space future without discrimination. We ask as citizens of this nation to be allowed to participate with seriousness and sincerity in the making of history now, as women have done in the past. There were women on the *Mayflower* and on the first wagon trains west, working alongside the men to forge new trails to new vistas. We ask for that opportunity in the pioneering of space."

She paused again. Now she had to convince them that women could begin training for space flight *now:* "There are sound medical reasons for using women as astronauts. Women weigh less and consume less food and oxygen than men, a very important point when every pound of humanity and the necessary life support systems is a grave obstacle in the cost and capability factors of manned space vehicles. Scientists say that women are less susceptible to monotony, loneliness, heat, cold, pain and noise than the opposite sex, vital facts to keep in mind in our nation's plans for space exploration of increasingly longer duration. . . .

"Now we who aspire to be women astronauts ask for the opportunity to bring glory to our nation by an American woman becoming first in all the world to make a space flight. No nation has yet sent a human female into space.

"We are thirteen human volunteers."

She smiled and said, "Thank you," and the audience began to applaud. They pushed back their chairs and stood, and the applause continued. When the meeting finally drew to a close a crowd of people surrounded the podium, and every one of them wanted to shake her hand and wish her well and express support.

By early summer NASA's efforts to recruit women personnel were proving successful. On June 27, James Webb announced that NASA employed 146 female technologists and 77 female mathematicians. But he said nothing about NASA's plans to hire women astronauts. There were no plans. Meanwhile, Jerrie Cobb had been an official consultant to NASA since June 1961, but she had never been consulted. It seemed that whatever NASA's motives regarding its other women employees, Jerrie's appointment had been a goodwill gesture that was, in the end, only a gesture. On June 30, her one-year appointment ended. It was not renewed.

26

THE HEARINGS

It seemed somehow meaningful that the space age and the nuclear age were coincident. During the same historical moment that humanity first began to tread the celestial realm, it also learned to destroy itself. Across America, people who had never seen the inside of a church or a synagogue were stirred to thoughts about larger meanings, thoughts about fate and eternity and the very purpose of human existence. The astronaut himself was a kind of emissary from the earth to the heavens. During Glenn's meeting with Congress somebody asked the astronaut about his religious faith. Glenn's answer: "My peace has been made with my maker for a number of years, so I had no particular worries on that line." That utterance received one of the biggest and longest ovations in the speech. It was as though through it all, there had been doubts—unspoken and almost embarrassed. And so to many, Glenn's answer was a relief. Whatever we were doing up there, our representative at least, was unblemished and pure.*

In the role of emissary to the heavens, John Glenn exceeded all

*In May 1962, in a comment widely reported, cosmonaut Gherman Titov told American reporters that during his seventeen orbits he'd seen no evidence of a deity.

expectations. In 1962, however, there remained other ideas about the proper role of the astronaut. Metaphysics aside, it was a rather complicated problem in a field that was coming to be called risk analysis. If the singular goal was to ensure that the astronaut returned alive, the odds were probably considerably improved by making sure that astronaut was a jet test pilot. Or, if the singular goal was scientific return, then that return would be insured by making the astronaut a scientist. But if the singular goal was to perform a "space first," then perhaps the benefits in national prestige outweighed the increased risks. The problem was that, in 1962, there was no singular goal. There was, rather, a collection of goals—among them, engineering advances, political prestige, scientific discoveries, and manifest destiny. Some were more important than others, and the importance of each seemed to change from week to week. And the whole problem was further complicated by the fact that even after two suborbital and two orbital flights, the risks were simply unquantifiable; so much was still unknown.

Even within NASA, questions about the role of the astronaut were still creating a good amount of consternation. To the engineers at McDonnell, the astronaut was a means to evaluate the performance of the spacecraft they had designed and constructed, in the environment they had constructed it for. This view was shared by the astronauts themselves, and by most members of the Space Task Group. For the first three Mercury flights, this position held sway. The missions of Shepard, Grissom and Glenn were primarily "engineering test flights," meant to evaluate the capabilities of the spacecraft and its systems.

As for the scientific community, by early 1962 American astronomers, meteorologists, and physicists had been allowed little say in Project Mercury. Although NASA had justified the expense of the program in part by allowing that the astronaut would act as scientist, and although the National Aeronautics and Space Act of 1958 had defined NASA's purpose as furthering "the expansion of human knowledge of phenomena in the atmosphere and space," there had been little real science performed on the flights of Shepard, Grissom, and Glenn for the simple reason that science experiments were considered a luxury (perhaps a dangerous one) when the spacecraft themselves were insufficiently tested.

So the scientists had been left behind, but not for lack of trying. As early as the fall of 1961, task group psychologist Robert Voas had asked NASA headquarters to consider having the astronaut carry out scientific experiments, and NASA headquarters had asked astronomer Jocelyn R. Gill to head an ad hoc committee overseeing astronomical tasks for the astronauts.* In the beginning, the recommendations were for rather modest experiments. But Glenn's mission had been so successful that planners saw an opportunity to put real science into Project Mercury. The next flight—it was Scott Carpenter's—was designed to yield as much scientific information as possible, so long as the proposed experiments were compact and lightweight, and in no way compromised safety. Again, Jocelyn Gill headed a committee charged with suggesting experiments. Along with other scientists—from the U.S. Weather Bureau, the Massachusetts Institute of Technology, Goddard Space Flight Center, and within NASA headquarters—she submitted proposals to the newly created Mercury Scientific Experiments Panel. Five simple experiments were selected.†

During Carpenter's flight, the science experiments produced modest successes, but the mission itself very nearly ended in disaster. Carpenter had begun reentry with nearly empty fuel tanks mostly because he'd wasted too much fuel moving the capsule unnecessarily (he accidentally actuated the high-thrust control jets six times), and so in the crucial moments before reentry he did not have enough fuel to properly align himself. It was only luck that he happened to fall heat shield first. Further, in part because he'd waited too long before hitting the retro switch, he overshot his designated landing zone.

NASA put a good face on all this. Its public position was that Carpenter had merely suffered fatigue, and that the flight was a qualified success. But a number of NASA staff, among them flight

*Gill was one of a handful of women who had reached the upper ranks of the young NASA hierarchy. Another was astronomer Nancy G. Roman, whom NASA had chosen in 1959 to direct the long-range development of astronomical observatories.

†Carpenter would release a colored balloon outside the spacecraft to measure atmospheric drag, he would observe the behavior of liquid in a weightless state inside a closed glass bottle, he would measure the visibility of a ground flare with a light meter, and he would photograph weather and the atmosphere's airglow layer. He would also use the small control thrusters to maneuver the spacecraft into a variety of orientations.

director Chris Kraft, remembered that at the time he was chosen as an astronaut, Carpenter had only 2,800 hours of any kind of flying, a mere 300 of that in jets, while some of the others had 2,000 hours in jets alone. Although Carpenter had graduated from the Navy's Test Pilot School at Patuxent, Maryland, he had done relatively little work afterward as a jet test pilot. The astronaut himself suspected that he had been chosen primarily because of the results of his examinations at Lovelace and of his psychological stress tests at Wright-Patterson—he had broken quite a few of their records. But after his flight there was, in the offices of NASA, some reevaluating. The lesson of Carpenter's flight seemed to be that the agency was pushing too hard. The Mercury spacecraft was still experimental. Experimental high-performance jet aircraft were not considered operational before hundreds of test flights; the Mercury/Atlas combination had had only four. Carpenter's flight had shown that it was too soon to give much attention to science and medicine. So if NASA administrators in 1962 were to rank the roles of the astronaut, they would have put engineering test pilot first, with scientist and medical test subject running distant seconds. As to national symbol, that was a consideration best left to politicians.

In the summer of 1962, the question of the proper role of the astronaut had a renewed urgency, because on April 18, NASA had begun accepting applications for Astronaut Group II—the astronauts who would fly the Gemini and Apollo programs. In the interim, NASA had reevaluated its criteria for astronaut candidates. Experience had shown that space flight was far less stressful on the human body than some had feared. So while candidates for Group II were required to be in excellent physical condition, the medical tests administered were neither as exhaustive nor as rigorous as those given the first group. Slight changes were also made in other requirements. Because the Gemini and Apollo spacecraft would be larger than Mercury, the height limit was raised to six feet even. The maximum age was lowered to thirty-five, because NASA would need the new group of astronauts to be flying missions well into the seventies. The college degree required could be in the biological sciences, and a letter of recommendation would be required from a supervisor or employer. Although the requirement that candidates be jet test pilots would stand (if anything, Gemini and Apollo would require more piloting), NASA would accept as candidates civilian jet test pilots—that is, nonmilitary pilots in the employ of an aircraft manufacturer or of NASA itself.* Opening the job to civilian applicants would close the ideological schism between a civilian agency and military employees. Perhaps more importantly, it would signal that NASA had every intention of making space flight as inclusive a venture as it could reasonably be.

The thirteen women who had proven themselves medically fit to become astronauts now saw a chance to qualify under the requirements of Group II. Although the new deemphasis on medical testing made the women's remarkable success at Lovelace considerably less significant, the inclusion of civilian test pilots meant that, theoretically, women could now meet the entire list of established criteria. And although by 1962 prejudices against women in the aircraft industry were widespread, NASA and private firms had no standing prohibitions against woman pilots. So the decision to include civilian jet test pilot experience in the criteria for an astronaut candidate made it possible for a woman to become an astronaut through established channels, at least theoretically. In truth, however, in 1962 only one American woman had significant experience in high performance jet aircraft. This, of course, was Jackie Cochran, and whatever her precise age, clearly it was above both the original upper limit of forty and the new one of thirty-five.

Meanwhile, Jerrie Cobb persisted in making a case for women astronauts. All she asked was a fair hearing. Suddenly, after months of letter writing and telephone calls, she was at last granted the opportunity. In June 1962, she and Jane Hart were approached by Representative George Miller, chair of the House Committee on Science and Astronautics. Miller was proposing a special subcommittee to examine the qualifications NASA required of astronaut candidates. In order that the issue be explored as thoroughly as possible, it seemed desirable to present the subcommittee with a test case. So he asked Jerrie and Jane if they would tell their story

*The relevant information would be entered into the transcript of the hearings: "Approximately 125 civilians could [in 1962] qualify as current, experimental test pilots, qualified in jets. Of these, 65 are employed by the Federal Aviation Agency or the National Aeronautics and Space Administration and the remainder in the aerospace industry."

to Congress. Formal hearings would, he explained, take place in mid-July, and they would last for two or three days.

Halfway through 1962 NASA was having a pretty good year. Things were at last starting to work, and if the Russians were still ahead in some areas, at least we were giving them a run for their kopecks. For NASA in 1962, even obtaining funding was easier. When the House Committee on Science and Astronautics had opened its annual budget hearings on February 27, 1962, its members expressed almost unanimous satisfaction with Project Mercury. And when administrator Webb requested $3.7 billion for fiscal year 1963, most of it was approved without question. So NASA's officials were a little bewildered when, a few months later, Congress announced that the agency would have to justify its criteria for selecting astronauts before a special House subcommittee.

The Committee approved of NASA's record on the whole, but it would have to answer to taxpayers, and that 3.7 billion required justification. Further, the letters of Jane Hart and Jerrie Cobb—and a number of letters and telegrams from other women—were having a cumulative effect. In 1962, when NASA reevaluated its original criteria for the selection of astronauts, Congress too gave this matter some thought. But Congress began its thinking from a rather different starting point. If the preponderance of opinion at NASA was that the astronaut was first and foremost an engineering test pilot, many members of Congress would have said that he was most importantly a national symbol.

Jerrie had spoken now to civic groups and flying clubs and women's organizations. A congressional subcommittee, though, was something else again. Jerrie had never had much confidence in her powers of persuasion, but Jane had arranged audiences with senators and congressmen and even the vice president, and had found in George Miller a more receptive ear than they had initially imagined. Jerrie thought that if Jane couldn't make Congress see the importance of sending women into space, then no one could. Jerrie knew that this might well be the last chance. If they couldn't persuade the members of the subcommittee, there would probably be no American woman in space for a long, long time.

NASA's position would be presented by George Low, currently

director of Spacecraft and Missions for the manned space flight program, and by two of the astronauts themselves—John Glenn and Scott Carpenter, the first and so far only Americans to orbit the earth. There was also to be another expert witness whose voice was likely to carry some weight, and this witness, Jerrie and Jane felt certain, would speak on their behalf. The subcommittee had invited testimony from Jackie Cochran.

Only Jerrie and Jane had been asked to participate in the hearings, so the eleven others stayed home and waited. Jerri Sloan, Irene Leverton, Bernice Steadman, and the others found it strange to be doing the kinds of things they did every day—taking the kids to school, flying a routine charter, fixing supper—knowing that at the same time in Washington their future was being discussed by members of Congress and NASA officials and even two of the astronauts themselves. It was a little frustrating, too, that the whole matter was out of their hands. But for the most part, they were optimistic. The very existence of a special subcommittee on the selection of astronauts was cause for hope. Jerrie and Jane, they knew, would present a strong case on their behalf.

To Jean Hixson back in Akron, though, there was something familiar about phrases like "special meeting of the House Subcommittee" and "testimony from expert witnesses." The last time such phrases had had this much significance in her own life, she'd been a WASP, listening to reports about Hap Arnold's testimony before another House Subcommittee, waiting for word on the future of the Costello Bill. Jean hoped that things might end differently this time. But she did not hope too much.

The first day of the hearings was July 17. The *Post* predicted clouds in the morning, followed by partial clearing. High about eighty. In other words, a perfectly normal day for the nation's capital in July. A nice day for flying.

At 10:00 A.M., in Room 356 of the Cannon Building, Jerrie and Jane were seated at a table, facing eleven members of the Special Subcommittee on the Selection of Astronauts. The subcommittee sat on a dais behind nameplates and microphones. There were nine men and two congresswomen—Corinne Riley of South Carolina and Jessica Weis of New York. Seated in the center of the dais was

Victor L. Anfuso, the subcommittee's chairman and the senior congressman from New York. He began the proceedings:

"Ladies and gentlemen, we meet this morning to consider the very important problem of determining to the satisfaction of the committee what are the basic qualifications required for the selection and training of astronauts. There is no question that our manned space flight program must make use of every available resource that can contribute to its success. As we look into the future, we can see greater and greater demands for special talents placed upon the people from whom future space travelers will be drawn. We are particularly concerned that the talents required should not be prejudged or prequalified by the fact that these talents happen to be possessed by men or women. Rather, we are deeply concerned that all human resources be utilized. . . ."

Anfuso named the speakers from whom the subcommittee would hear that morning: "Miss Jerrie Cobb, a noted pilot with an outstanding career in aviation; Mrs. Philip Hart, wife of Senator Philip A. Hart, of Michigan, and also a famed pilot, as well as an outstanding wife and mother; and later Miss Jacqueline Cochran, of whom little more need be said than that she holds more national and international distance, speed, and altitude records than any other pilot."

He went on. "We will hear your prepared statement, Miss Cobb. And we are very happy to have you here. We know the great effort you have made, together with the other twelve women astronauts, in calling the subcommittee's attention to this particular phase of the astronaut training program. . . . This committee is noncommittal, of course, but we will make sure that all the facts will be heard and properly presented to the Congress of the United States. You may proceed."

As he yielded the floor, Jerrie stood, and wished she weren't feeling quite so shaky. She looked at the dais, managed a quick smile, glanced down briefly at her notes, and began:

"Thank you, Mr. Chairman, honorable Members of Congress. May I say first, on behalf of myself and the twelve other women space candidates, that we thank you for the opportunity of letting us be heard on Capitol Hill. We appreciate the vision and interest you are showing in recognizing the need for looking into the utilization of women in the U.S. space program on a serious, sound basis. . . . Our purpose in appearing before you is single and simple: We hope that you ladies and gentlemen will after these hearings and due consideration, help implement the inclusion of qualified women in the U.S. manned space program."

She began to explain the logic behind this request—how Lovelace and Flickinger had begun putting women pilots through the Mercury tests; how fully half the women tested had met the medical qualifications for space flight; how the tests had shown women to be as suited for space travel as men, and in some respects better suited; and how sending a woman into space could be a valuable first for the United States. She asked that the qualifications and brief biographical data on the other eleven women who had passed the tests at Lovelace be entered into the Congressional record. And she concluded:

"Members of Congress, your special subcommittee sits here today in search of the 'practicability' of training and using women as astronauts. It is clear to us, and we hope to you, that the practicability exists and is at hand. We welcome your questions."

Anfuso cleared his throat. "Miss Cobb, that was an excellent statement. I think that we can safely say at this time that the whole purpose of space exploration is to some day colonize these other planets, and I don't see how we can do that without women." The remark drew general laughter. Anfuso smiled and turned the floor over to Jane Hart.

Jane could give as good as she got, and she began, "I couldn't help but notice that you call upon me immediately after you referred to colonizing space." Again there was laughter. But if Jane Hart knew how to crack a joke, she also knew how to stir the imaginations of politicians, and she knew how to appeal to their sense of obligation to the citizens they represented. She sought to convince them that the objectives of a woman in space program were also the objectives of Congress.

"I am not arguing that women be admitted to space merely so that they won't feel discriminated against. I am arguing that they be admitted because they have a very real contribution to make. Now, no woman can get up and seriously discuss a subject like this without being painfully aware that her talk is going to inspire a lot of condescending little smiles and mildly humorous winks. But happily for the Nation, there have always been men, men like the members of this committee, who have helped women succeed in roles that they were previously thought incapable of handling.

"A hundred years ago, it was quite inconceivable that women should serve as hospital attendants. Their essentially frail and emotional structure, it was argued, could never stand the horrors of a military dressing station. Well . . . the women were insistent. There was a shortage of men to do the job. And finally it was agreed to allow some women to try it. . . .

"I submit, Mr. Chairman, that a woman in space today is no more preposterous than a woman in a field hospital 100 years ago. And I further submit that the venture would be equally successful. . . .

"I wonder if anyone has ever reflected on the great waste of talent resulting from the belated recognition of women's ability to heal. Before 1862, there must have been thousands of women with innate nursing ability who might have saved countless lives if only they had been allowed to. But in this scientific field, no one recognized what women could do because they were never permitted to try. It seems to me a basic error in American thought that the only time women are allowed to make a full contribution to a better nation is when there is a manpower shortage."

Jane spoke of letters from young women science students discouraged at their prospects. She noted that there were fewer graduates in engineering of either gender than there had been ten years earlier, while the Soviets were educating three engineers for every one in America. We had fallen behind, she said, and we had only ourselves to blame. She was ready with one part of the solution: "Now, I think that women should be allowed to go into space without delay." It was a strong position, unpalatable to many, and she also offered a moderate stance: "But even the extreme view that women will have no place in outer space for many years does not justify the cancellation of a research program that has already begun and that would doubtlessly supply information useful right now as well as in the future. Actually, the reinstatement of this research program will have a dual purpose. First, it will furnish valuable data. Second, it will encourage more talented young women to enter the specialized fields relating to space engineering. I think it would open to the nation a great new reservoir of ability and enthusiasm.

"Above all, I don't want to downgrade the feminine role of wife, mother, and homemaker. It is a tremendously fulfilling role. But I don't think, either, that it is unwomanly to be intelligent, to be courageous, to be energetic, to be anxious to contribute to human knowledge. I just think we would be making a serious mistake if we assumed that women just have no contributions to make to space exploration. I just think we would be making a serious mistake if we were not willing to at least research the *possibility* that they could make a contribution. Thank you."

The floor was opened for questions, and Anfuso began: "May I ask you, Mrs. Hart, whether your flying activity has encouraged other people, both male and female to take up flying?"

Jane said, "Yes, sir. I could give you some very specific instances of when I have taken people up who would not even set foot in an airliner and who subsequently have taken up flying as either a hobby or a career."

Anfuso asked, "Would you also say that if you had the opportunity to participate in the manned space flight program, your experience would encourage other women to take up science and engineering courses, particularly those applicable to space travel?" Again Jane answered in the affirmative. This wasn't as difficult as Jerrie had feared. Anfuso asked for more about the nature of the tests.

Jerrie had prepared a set of slides depicting the three phases of testing, so the room was darkened and an image of Dr. Shurley's isolation test tank appeared on a large screen. The projector hummed and clicked and there was an image of Jerrie herself in the tank taken in infrared light. There were several more slides, and Jerrie explained each. After the last, the lights in the room brightened. Anfuso asked, "Miss Cobb, what do you think are the minimum qualifications for an astronaut?"

She paused a moment, then said, "I could not answer the minimum qualification for an astronaut because I am not qualified . . . The qualifications that the authorities of NASA have set down have made it impossible for women to qualify as astronauts or even demonstrate their capabilities for space flight . . . It is the jet test pilot experience that makes it impossible for a woman to meet the qualifications . . . Some of us have worked as test pilots, but it is impossible for a woman in this country to be a jet test pilot because there are no women pilots in the military services and the test pilot schools are operated solely by the military services."*

Congressman Joseph Karth of Minnesota asked Jerrie if she thought it essential that an astronaut be a test pilot, and Jerrie replied that in her opinion it was not. Further, she said, the women who passed the tests at Lovelace had logged an average of 4,500 hours of flying time—more than the average held by the astronauts. To this, Karth suggested that there was "considerable difference between straight flying—commercial or private—and test piloting." Jerrie had heard this objection before, and she was prepared: "I suggest there is an 'equivalent experience' in flying that may be even more important in piloting a spacecraft. Pilots with thousands of hours of flying time would not have lived so long without coping with emergencies calling for microsecond reactions. What counts is flawless judgment, fast reaction, and the ability to transmit this to the proper control of the craft. We would not have flown all these years . . . without accumulating this experience. This experience is the same as that acquired in jet test piloting. I think you might acquire it faster as a jet test pilot, but it is by no means the only way to acquire it."

There ensued a discussion between Jerrie and Karth of the flying hours required for astronauts and test pilots. Anfuso broke in and said, "May I interrupt to welcome Miss Jacqueline Cochran, who has just arrived?"

In 1962 the lines around Jackie Cochran's eyes were too deep for even the heaviest foundation creams to conceal, but she was still an imposing figure in her dark, smartly tailored suit, and she was still the

*The nation's two test pilot schools were the Naval Air Test Center at Patuxent River, Maryland, and the Air Force Experimental Flight Test School at Edwards Air Force Base in California. Technically, Cobb was wrong here. Most civilian jet test pilots had learned to fly jets in the military, but there were exceptions: An X-15 pilot named Joe Walker, for instance, had flown his first jets as a civilian employee of NACA. In 1962 no hiring policies made it impossible for a woman to become a jet test pilot without graduating from a test pilot school, although widespread prejudices would have made it extremely difficult. In 1962 only one American woman had acted as a test pilot for new or current high performance jet aircraft: Jackie Cochran had flown for North American in 1953 and for Northrop during the previous summer and fall. As noted earlier, her employment in both cases was the result of an unusual or unique set of circumstances.

It is interesting that World War II had offered women unprecedented opportunities not only in military but in civilian piloting. Barbara Jayne, Elizabeth Hooker, and Cecile "Teddy" Kenyon had worked as production test pilots for Grumman, piloting Avengers, Wildcats, and Hellcats on initial flights.

same Jackie Cochran who lived her life according to her own schedule, who thought nothing of showing up late for a meeting with members of Congress, who simply strolled in during the middle of the proceedings and with such aplomb that the subcommittee chairman saw fit to bring everything to a halt so everyone present could acknowledge her arrival. She made her way across the room, smiling at an old acquaintance here and there. Jerrie and Jane were glad that she had come. She would, they were certain, be a valuable ally.

But before Cochran gave her testimony, there were more questions. Congressman James Fulton of Pennsylvania was obviously sold on the notion of women astronauts, and wanted to help Jerrie overcome Karth's reservations. Now, he asked, who made safer pilots, women or men? No one in the room had a precise answer, so Fulton asked if it were not true that women were safer automobile drivers than men. When Jane Hart admitted that as far as she knew the statistics bore this out, Fulton pronounced himself satisfied that better drivers would make better astronauts. In fact, he said, "the problem is not whether you are equal to men. To me it would be: How superior are women over men?"

Jane wasn't sure how to respond. She appreciated his support, but recalling that this was the same Congressman Fulton who had advocated blue dust on the moon, she thought it best to pass on that particular line of inquiry. Jerrie and Jane tried to return to more immediate matters, and Anfuso asked for clarification of their positions. They replied that although they were not experts, in their opinion NASA's idea of the necessary background for an astronaut was unreasonably narrow—the astronaut need be neither test pilot nor engineer. Anfuso asked Jerrie and Jane whether they thought the United States should be the first nation to launch a woman astronaut. Their answer, as everyone in the room must have expected, was in the affirmative. Anfuso had an immediate follow-up: "Do you believe that the recognized hazards in such a feat and the possible worldwide repercussions to our prestige in the event of a tragic accident, are worth the risk and expense for us to achieve that objective?"

It was a good question. The recent successes of Glenn and Carpenter made it all look easy. In truth, though, NASA had never been entirely confident, especially in the first months. They had assigned a committee called the "Mission Analysis Branch" to act as professional pessimists, considering what might go wrong. Indeed, there was much that *could* go wrong. Eighty-thousand parts on the Atlas/Mercury were termed "critical"—that is, their failure could end the mission, the astronaut's life, or both. A dead astronaut was a very real fear, and it was present at the highest levels. Before Shepard's flight, Kennedy had made last-minute phone calls to ask questions about the escape tower, and to be assured that he had not approved a multimillion-dollar government-sponsored funeral pyre, funded by taxpayers and broadcast live on national television. As bad as that would be, Americans were used to men dying in combat, and because the astronaut was a cold warrior, his death would be an extension of the known. But a woman? Somehow that would make it worse. So Anfuso's question was a polite way of asking, *do you think this is worth sacrificing your life?*

Jerrie said, "I very definitely do."

Jerrie elaborated her answer, alluding briefly to NASA's testing of chimpanzees for space flight. When Congressman R. Walter Riehlman asked Jerrie if she were willing to undergo similar testing, she said, "Even if I have to substitute for a female chimpanzee." The questions continued. Jerrie and Jane answered as best they could, and the same questions were repeated, and one or the other of them would have to explain the same point yet again.

Then Congressman J. Edward Roush said, "Miss Cobb, I couldn't help but overhear a conversation between you and Mr. Anfuso prior to the hearing, and during the course of the conversation you said, 'I am scared to death.' How do you reconcile this emotional statement with the fact that an astronaut must be fearless and courageous and emotionally stable?" She smiled wryly and said, "Going up into space couldn't be near as frightening as sitting here." Again there was laughter, and Anfuso said, "I think, in conclusion, I might say, Miss Cobb, that what you would like to accomplish is a parallel program, but not to interfere with any existing program; is that correct?" And Jerrie said, "I think it need not be a separate program, nor interfere with the current program." There was a short recess, and the committee prepared to hear from Cochran.

Jackie Cochran had a prepared statement, and it wasn't quite what Jerrie and Jane had expected. "I do not believe there has been any intentional or actual discrimination against women in the astronaut program to date . . . The determination of whether women should be included at this time in the program of training astronauts should not depend on the question of sex but on whether such inclusion will speed up, slow down, make more expensive, or complicate the schedule of exploratory space flights our country has undertaken . . . I believe, based on my experience with women in the WASP program, that women will prove to be as fit as men, physically and physiologically, for space flying. But such proof is presently lacking. It should not be searched for by injecting women into the middle of an important and expensive astronaut program. There is a simpler and sounder way available." Cochran then proposed an extensive long-term program to research women's fitness to serve as astronaut candidates.

Jerrie and Jane did not understand. More medical research? Hadn't they been poked and prodded at least as much as any human beings in medical history? What could possibly be accomplished by more research? They listened as Cochran outlined a program that would study hundreds of women, maybe for years, and they thought, but didn't we already establish that women were physically capable of space flight? Aren't we here today to discuss the other criteria for the selection of astronauts, the ones that prevent women from being considered? Cochran had her opinions of those criteria too, and for the next forty minutes they were revealed in the question-and-answer session. Did Miss Cochran believe that the requirement that astronauts be test pilots and engineers was sound? Yes, on the whole she did. But she would not go so far as to say either requirement was absolutely necessary. She reminded the committee that the highest degree held by Chuck Yeager was a high school diploma, and that John Glenn himself did not have an engineering degree.* Anfuso asked whether she thought it a worthwhile national objective to send a woman into space before the Soviet Union. Jackie said, "Well sir, that is a very difficult question to answer. I think the national objective would be to try to surpass them in every field of space exploration . . . but I think there are many things more important, let's put

*In fact, the Space Task Group's original seven-item list had required a degree in science or engineering "or its equivalent thereof." Two of the original seven, Carpenter and Glenn, had not attained the baccalaureate, but were judged to have gained comparable experience as test pilots or in test pilot school. Excepting Sarah Gorelick, who had a B.S. in mathematics, the women who passed the tests at Lovelace failed to meet the Space Task Group's original requirements not only because they were not jet test pilots, but because they had no degree in science or engineering, although several may have been able to make a viable case for its equivalent.

it that way. Sure it is nice to be first, but it is also nice to be sure . . . I heard more than a year ago . . . not with any authority but through the usual scuttlebutt [that Russia was going to] orbit a woman. In the meantime they have orbited two men. The gossip is that they have had some difficulty and it may be sometime before they orbit another man." Congresswoman Weis asked her whether she would support the completion of the program of tests Lovelace had begun on Jerrie and Jane and the others. Cochran replied, "I donated the funds for these women to have the medical examinations . . . If I had not been interested . . . I would not have gone to this expense and trouble."

Cochran's testimony was not without its more human moments. Congressman Fulton, proposing that NASA consider using women as crewmembers, offered a list of influential women throughout history—Molly Pitcher, Sacajawea, Queen Elizabeth I. When he reminded the committee that Pocahontas offered her life so that Captain John Smith could live, Cochran grinned, and said, "I think that was done for love, sir. Women will do an awful lot for that."

In sum, Jackie Cochran believed that sending women into space was not worth diverting any funds or energies from the program in progress, and that such decisions were best left to NASA itself. Her final word on the question of an American woman in space was, "I would like to see us do it properly and successfully rather than to make a mess of it."

They adjourned at noon. Jerrie and Jane's active role in the proceedings was over. On the whole they felt optimistic about their reception. They puzzled a bit over Cochran's remarks, but there seemed some room for compromise between Cochran's view and Jane Hart's moderate position. At any rate, if Cochran was not helping their cause, her very presence lent it credibility. Some of the committee's members, most notably congressmen Fulton, Hechler, and Corman, were obviously sympathetic. Jerrie and Jane walked outside and down the steps. They saw that the *Post* had been right. The sky had cleared. A good day for flying. Ceiling and visibility unlimited.

On July 18, the meeting of the Special Subcommittee on the Selection of Astronauts again came to order at 10:00 A.M. On the dais were the eleven members of the subcommittee as well as George Miller, chair of the Full Committee. Sitting at the table before the dais were astronauts John Glenn and Scott Carpenter, and director of Spacecraft and Flight Missions George Low. The room was a bit more crowded; in the visitors gallery at the back were several mothers who had brought their children to see the astronauts.

Chairman Anfuso introduced George Low. Low had been an influential member of the Space Task Group that had first determined astronaut qualifications, a member of the NASA–ARPA Manned Satellite Panel, and an engineer at the NACA Lewis Flight Propulsion Laboratory. Within NASA itself, Low was known as the author of NASA's proposal to launch a spacecraft around the moon. He had been director of Spacecraft and Flight Missions since November 1961.

Low was neither politician nor public relations specialist, and now he was expected to act both parts. He began by reminding the committee that NASA was selecting the next group of astronauts, and that the original selection criteria had already been reevaluated. He spent some time delineating the rationale behind the jet test pilot requirement debated on the previous day:

"In many ways, manned spacecraft can be considered the next generation of very high performance jet aircraft. Their velocity and altitude capabilities are very great. A spacecraft has life-support systems, control systems, landing systems, power and fuel systems, and many other similarities with high-performance jet aircraft. Thus, there is a logical reason for selecting jet test pilots."

Then Anfuso asked Glenn, "Do you think that an astronaut or crewmember necessarily has to be a test pilot?" It had been almost three years since the astronauts had won the battle to act as test pilots. They had succeeded in persuading the capsule's designers to create a re-arranged control panel, an escape hatch, a window, and an emergency manual control system; they had even persuaded NASA to call the capsule a "spacecraft" in technical documents and press releases. So today they were fighting a battle they thought they'd already won. Glenn began, "Let me preface my remarks by one statement. I am not 'anti' any particular group. I am just pro space . . . . Anything I say is toward the purpose of getting the most qualified people of whatever sex, color, creed, or anything else they might happen to be.

"The demands of just understanding the space vehicle systems require a good technical background. It is an experimental program. . . . The astronaut's function is actually . . . to take over full control, to analyze, assess, and report the various things that he encounters, or new situations in which he finds himself. In doing this he must perform these functions under periods of high stress, both mentally and physically, and observe many complex functions under these stresses . . . The person who can best perform all these functions is still represented most nearly by the test pilot background."

Representative Miller and Glenn discussed the duration of training for a test pilot, and Miller asked both astronauts to expound upon the reasons for their choice of professions. Low assured the members that the qualifications were not designed to exclude women. Anfuso asked Low whether a Russian woman in training would alter his position, and Low replied that it would not. Low added that a separate training program for women would be impractical because the training equipment—the centrifuge, vacuum chambers, and capsule procedures trainers—was in constant use.

Representative Karth introduced the position of Jerrie and Jane, saying, "Yesterday, two of the witnesses spoke very strongly about [the test pilot] qualification. They felt, quite frankly, that an extensive number of logged hours in actual flight compensated for all the variables and invariables and the emergencies that one might meet as a test pilot."

And Low said, "I think I will have to disagree . . . It is true that other pilots also get into stressful situations, but not as often or as frequently as these men do . . . We all know that in John's flight, he had trouble with his automatic control system. He had to assess the difficulties and then calmly go on the manual system and use it, and use it effectively, under trying conditions. I am also sure that the decision to leave the retropack on in John's flight would have upset anyone, man or woman, that did not have the kind of training that these men have had."*

Carpenter added: "A person can't enter a backstroke swimming race and by swimming twice the distance in a crawl qualify as a backstroker . . . [A] preponderance of hours received in normal ci-

*Glenn's automatic control system worked only intermittently, and he flew much of the mission in "fly-by-wire" mode. Near the end of his first orbit, ground crews received a

vilian flying does not compensate for a lack of military jet test flying." Glenn agreed. "To say that a person can float around in light planes or transports for—I don't care how many thousands of hours you name—and run into the same type of emergencies that he is asked to cope with in just a normal six-month or one-year tour in test flying is not being realistic." He added, "Instead of trying to reduce our qualifications to a lower level . . . perhaps we should be upping the qualifications and saying that we have to have test pilots with doctorate degrees and with even more experience than we have had to date so far."

Fulton had been listening patiently. Now he put his position in clear terms: "If a woman . . . through her experience and her flight experience can give the equivalent capabilities and characteristics for a good astronaut, she should not be rejected because of a requirement that she is unable to fulfill." Now Fulton had the floor, and spoke at some length, continuing, "I believe that the United States should adopt a program to put the first woman in space. We should set that as a national goal. . . ." Glenn and Carpenter and Low and even the members of the subcommittee seemed increasingly uncomfortable. Finally the congressman from Pennsylvania had finished, and he asked Glenn: "Would you agree?" The astronaut said, "I think this is a little out of my province, sir." There was laughter, and Glenn said, "I am not qualified to judge what national goals we should have in this regard, sir."

Fulton suggested that NASA's evident desire to protect women also disallowed them from participating in the human adventure. Again he listed women who influenced the course of history—Molly Pitcher, Sacajawea, Queen Elizabeth I—then asked, "Don't you think that they could pass those tests as well as men?" The prospect of Queen Elizabeth and Sacajawea as guests of the Naval School of Aviation Medicine in Pensacola had an absurd quality, appreciated by the audience in general and by astronaut Glenn in particular, who answered with mock solemnity, "Oh, yes sir." There was quiet laughter, but Fulton pressed harder: "You must

---

signal that indicated that the landing bag between the capsule's heat shield and inner bulkhead had inflated, a condition that meant that the heat shield might be loose. Consequently the retropack, normally jettisoned before reentry, was left strapped to the heat shield as a means of holding it in place. The signal turned out to have been false.

remember that Ham made a successful trip, too . . . I think a woman could do better than Ham."

Several postflight reports of the missions of Ham and Enos had noted that problems in both would have been corrected had a human pilot been aboard, but the fact remained that one chimpanzee had ridden a Mercury capsule into space out over the Atlantic, and another had ridden another Mercury for two orbits around the earth. Fulton's allusion to Ham was an unintentional slur against the four brave American men launched into space, and probably the congressman did not know how close he had come to the bone. Glenn reddened a bit, and said "That is not a fair comparison, sir, with all due respect." Then the astronaut leaned forward and said, "I would like to point out too, that, with all due respect to the women that you mentioned in all of these historic events, where they performed so fine, they rose to the occasion and demonstrated that at the time they had better qualifications than the men around them, and if we can find any women that demonstrate that they have better qualifications for going into a program than we have going into that program, we would welcome them . . . *with open arms*." This got a roar from the crowd, and Glenn sat back in his chair and looked around as though surprised. Then he smiled a little, and when the laughter stopped he leaned into the microphone again and said a little more quietly, "For the purposes of my going home this afternoon, I hope that will be stricken from the record." Again there was general laughter.

For several minutes there was a discussion of a possible parallel program for women, with assurances by Low that qualifications would be reevaluated for the selection of Group III and that there had been no deliberate discrimination against women. Anfuso said, "I am sure that if we had lower standards than those that you have outlined, it might be dangerous. I think the loss of prestige in losing a woman in space would certainly be something we would hear about. So let us not be too hasty in changing those regulations." Then, speaking for the subcommittee, he urged Low to consider a parallel testing and/or training program for women.

Low reiterated that such a program would inevitably impede the pace of the present program. Congresswoman Weis asked Low what he knew of the tests at Pensacola. NASA, he said, had not requested the tests in the first place, and he could explain neither the cause for their scheduling nor the reason for their cancellation. Weis asked whether anything might be done to enable women to be accepted as test pilots. Low said he imagined that the dearth of women test pilots probably was a product not of discrimination, but of the fact that so few women chose the field to begin with. Glenn called the low numbers, "a fact of our social order."

Congressman Joe E. Waggoner of Louisiana, quiet until now, said, "I don't think that anyone can deny that the criteria which NASA has used to this point in selecting astronauts has been one which has been successful and is beyond criticism, as far as I am concerned." Congressman Walter Moeller of Ohio noted that many industrial employers reported that women cost more to employ than men, and suggested that training women would cost NASA more than training men. Then Anfuso said, "I think we are all in accord on this committee that we should do nothing to interfere with the present NASA program . . ." Anfuso seemed not to have heard the comments of Waggoner and Moeller, because he continued, "I think that it is also the general feeling of this committee that this country is big enough to tackle a program for women. In line with what Miss Cochran said, she had a program, and a good program, I think." Low agreed that the nation should not exclude women from any activity, yet again expressed doubts about a program to train women in that it would slow the present program. Then he turned Anfuso's argument on its head: "[I]f we *don't* relax our requirements . . . this in itself I think would be an impetus for more women to get into the field of engineering and science, which will be of benefit to our total space program."

Congressman Fulton offered a brief history of American women in aviation, and took issue with Low's claim that NASA, recently allocated three billion dollars for various programs, lacked the funds to begin a small program for women. Anfuso interrupted that the House was in session, and the subcommittee would have to recess. He allowed Fulton a final question, and the congressman put it to Glenn: "You wouldn't oppose the proposal of the chairman and myself to have a small training program for women?" The astronaut replied, "I wouldn't oppose it. I see no requirement for it."

The hearings ended there, at an awkward impasse. What of substance had resulted? George Low had received a vague directive from Anfuso to consider a training program for women and to encourage the test pilot schools to admit women. But Anfuso was a congressman from New York, and Low was director of Spacecraft and Flight Missions. Neither was in a position to set NASA policy. And George Low, a middle-rank NASA administrator, could do nothing at all about entrance requirements for test pilot schools of the Air Force and Navy, even if he wanted to. To an observer in the visitors gallery, it might have seemed as though Low and Glenn and Carpenter had been made aware of matters about which they'd given little or no thought, and that perhaps a few seeds for the cause of women astronauts had been planted. But anyone working within NASA would have known that in the summer of 1962 the agency had little inclination to begin a new program, even a small one; NASA was at full throttle, gearing up for a lunar landing in eight years time.

The hearings were adjourned at 12:05 P.M. Jerrie Cobb and Jane Hart walked slowly down the steps of the Cannon Building into the bright noonday sun. There wasn't much to say. There were a few photographers waiting, and Jerrie and Jane made brave Trojan women smiles and said a few things into the microphones about fighting on.*

In their homes in Dallas and Kansas City and Akron, the others heard the news, and they were disheartened. But Jerrie insisted that their fight would continue. She urged them to keep writing to their congressmen, and to ask their friends to do the same, and they did.

*Because Jerrie and Jane had been given to understand the hearings would last for three days, and she was denied summation time, Jerrie submitted a "supplemental statement" for the congressional record. She pointed again to what she called built-in discrimination against women in NASA's selection process, and argued again for the validity of equivalent experience. She also suggested that the positions of Glenn, Carpenter, and Low were pure opinion, and observed that no expert—that is, no specialist in aerospace medicine—had been called upon to testify during the two days of hearings. She concluded that the argument that a training program for women would impede training of men, or the lunar landing goal, was wrong, and again asked that the women be granted an opportunity to prove themselves.

Cochran also submitted a supplemental statement. She observed that despite press accounts, there was no woman astronaut program, that the tests at Albuquerque were merely preliminary medical tests. She also suggested that Cobb had not been appointed spokesperson for the others, and indeed submitted a letter from Gene Nora Stumbough that expressed a view similar to her own.

For a while there were letters from people all over the country, wishing them well. But life had to go on. There were charters to fly and suppers to cook. Every now and then there was a free moment which might be spent looking upward, maybe noticing a tiny point of light moving slowly across the evening sky.

Through the rest of the summer, Jerrie wrote more letters and made more telephone calls and gave still more talks to civic groups and women's groups and pilots' associations. She and Jane appeared on the *Today* show and on radio programs, and in September, *Life* magazine named Jerrie of the one hundred most influential Americans under the age of forty, calling her "a likely first among the thirteen U.S. women astronaut candidates." It seemed the journal was unaware that in September 1962 the United States had *no* women astronaut candidates.

When the women talked about the hearings, Jackie's testimony puzzled them most of all. Why had she come to their assistance and then argued against letting them go on? On the surface, her stance made no sense. But they tried to make sense of it, and in the end there were several views. Jerri Sloan, for one, was not a bit surprised. Jackie was getting old, and the old are always envious of the young. Jerri suspected that when Lovelace told Jackie she would never pass the medical tests, Jackie knew she would not be the first woman in space, and wanted to make sure no other woman would go either. Some of the others suggested she'd been after a chance to relive her glory days: Her proposal for an extensive medical testing program involving hundreds of women had a familiar ring. Bernice Steadman admired Jackie, and thought the boys at Edwards just got to her. She flew with them and talked with them, so it was only natural that their way of looking at things would have rubbed off. Irene Leverton would later say that Jackie, in her heart of hearts, knew that NASA was not about to support a program for the long-term, so she put an end to it before it started.

27

THE MISSILES OF OCTOBER

One reason Jerri Sloan was unable to attend the hearings in Washington was her work. In the summer of 1962, she was co-owner, administrator and chief security officer for Air Services. Air Services pilots were test-flying the first infrared imaging system for Texas Instruments. Because Texas Instruments was under contract with the government, the work was top-secret. A tractor would pull one of Air Services' B-25s a few hundred feet to the Texas Instruments hangar. There were guards everywhere—inside and outside the fence, at every door. No one could get in or out of that hangar without a security check, and no cameras were allowed inside. The tractor would pull the B-25 inside, the hangar door would close, and the engineers would install the equipment. Then the hangar door would open and Air Services pilots—sometimes Joe or Jerri—would fly the plane. They flew over east Texas, they flew over west Texas. Sometimes they flew over downtown Dallas. The engineers, riding in the midsection, took their pictures. The B-25 would return to the hangar and the crews would remove the equipment and the Texas Instruments engineers would recalibrate, reload film, and then Air Services pilots would fly it again. In 1962 it went on continually, and Air Services pilots were on call twenty-four hours a day.

The system was amazing. It could take photographs in the dead of night. With infrared cameras aimed through portholes in the fuselage from an altitude of 60,000 feet it could photograph a strip of earth 125 miles wide and 3,000 miles long. Air Services' B-25s and DC-3s could not flight test at such altitudes, of course. Still, the detail was beyond belief: they could see an elevator shaft by the heat it gave off. When Air Services pilots saw photographs taken with what they were flying, sometimes it was hard not to say something. There in the ready room, they'd sip some coffee and remark, "Well, nobody's gonna have any secrets anymore."

For Jerri herself, it was a strange life. In the morning she'd run security checks on job applicants, make flight assignments, and see about maintenance. By late afternoon she'd get in the station wagon and pick up David at Little League practice, and drive Candy to ballet lessons. For someone accustomed to speaking her mind, it was hard to know what to say to other mothers whose mornings had been full of laundry and groceries. And it was increasingly difficult to maintain friendships even with other women pilots. Jerri had been president of the Dallas chapter of the Ninety-Nines, and she still attended a meeting now and again. But it was hard. They could tell her about deliveries and instructing and demonstration flying for small aircraft firms. And then, when they'd ask Jerri what she was up to, she'd have to say, "I can't tell you."

On August 10, CIA director John McCone examined CIA reports on the movement of cargo ships from the Black and Baltic seas to Cuba. He dictated a memorandum to the president. He said that Soviet medium-range ballistic missiles were headed for Cuba.

A few days later the Soviet news agency Tass announced the successful launches of *Vostok 3* and *Vostok 4*. Americans had come to expect it by now. Everything the Russians did in space was spectacular, everything was another giant step. *Two men in space*. Two *Vostoks* placed into orbits so similar that they actually passed within three miles of each other. Americans did not have the facilities or systems reliability to coordinate simultaneous space flight. It took months to mate the capsule to the Atlas, to check its systems, to coordinate tracking and telemetry, and problems with any of these could disrupt the schedule. Now the Russians had systems so reliable that they had managed to get two off at the same time. In the U.S., editorials said it was obvious the Russians had launch facilities far larger and more sophisticated than ours. It was also obvious that the Russians were not far from being able to approach an American satellite and disable it. Soon they would construct space stations on which they would station soldiers from the Red Army. And the high ground would be theirs.

Jerrie Cobb was back in Oklahoma City when she'd heard the news of the Russian's twin flight. Her first thought was that it might encourage NASA to reconsider its position on women in space. As far as she was concerned, the outcome of the hearings was inconclusive, and she believed that among certain groups her cause was still gaining credibility. Webb himself had written her, expressing willingness to discuss the matter further. She in turn had written again to Webb, asking for time on the simulators when nobody was using them, and offering to take engineering courses. Jerrie knew that in the coming year there were to be two or three more flights in the Mercury program, and a woman might yet be inserted somewhere into that series. Or perhaps with Jim Webb's approval, the program, now scheduled to end in the summer of 1963, might be extended by one flight—an American woman in orbit as a fine grace note in the overture of the nation's first venture spaceward. It was still possible. On August 13, from Oklahoma City, she sent a telegram to the White House. It read: HOPE IN LIGHT OF NEW RUSSIAN SPACE ACHIEVEMENT YOU WILL RECONSIDER MY REQUEST TO DISCUSS WITH YOU UNITED STATES PUTTING FIRST WOMAN IN SPACE.

Probably Kennedy never saw that telegram. Soon enough he would be concerned with other matters. On August 29, a high-altitude U-2 surveillance flight provided conclusive evidence of the existence of SA-2 surface-to-air missiles at various locations in Cuba. On October 15, quick readout teams at the National Photographic Interpretation Center in Washington analyzed photos taken by a U-2 mission the day before. Late in the afternoon, one of the teams found the main components of a Soviet medium-range ballistic missile in a field at San Cristóbal.

Kennedy planned to address the nation on Monday, October 22.

The speech would be at 7:00 P.M. Eastern, 6:00 P.M. Central—and 6:00 in Dallas. In her living room Jerri Sloan watched the televised image of the presidential seal fade to Kennedy at his Oval Office desk. He was already looking straight at the camera, and there were dark circles under his eyes. He began immediately, speaking almost without inflection: "Good evening, my fellow citizens. The Government, as promised, has maintained the closest surveillance of the Soviet military buildup on the island of Cuba. Within the past week unmistakable evidence has established the fact that a series of offensive missile sites is now in preparation on that imprisoned island. The purpose of these bases can be none other than to provide a nuclear strike capability against the Western Hemisphere." The Russians had missiles in Cuba, and they could reach much of the southeastern United States. They could reach Dallas.

But there was, after all, work tomorrow. So like most people, Jerri put the children to bed. Then she went to bed herself.

Jerri Sloan walked into the hangar the next day, a little tired. The chief mechanic was there, and she said, "What do you think?" He looked at her, unusually serious. "I think if they don't pull 'em outa there we'll just hafta save 'em the trouble." Jerri said, "I think so too." All that week across the United States there was talk of shelters, and dark jokes, and difficult and protracted explanations for the children.

On Friday, October 26, Kennedy agreed to apply further pressure by increasing the frequency of low-level flights over Cuba from twice per day to once every two hours. The next day, around 12:00 noon, a U-2 reconnaissance plane was shot down, and its pilot was killed. It got worse. Sunday morning, the CIA's daily update as of 6:00 A.M. reported that Soviet technicians had made fully operational all twenty-four missile sites in Cuba. One nuclear bunker had been completed. Then, a new message from Khrushchev was broadcast on Radio Moscow: "The Soviet government, in addition to previously issued instructions on the cessation of further work at the building sites for the weapons, has issued a new order on the dismantling of the weapons which you describe as 'offensive,' and their crating and return to the Soviet Union." On November 1, photoreconnaissance showed that all missile sites had been bulldozed and that the missiles and associated launch equipment had been removed. Construction appeared to have stopped, and the installations were partially dismantled. As quickly as it had begun, it was over.

For several days the world had contemplated nuclear war. By the first months of 1963 it seemed that everything about the international situation had changed. It also seemed that nothing had changed. At Air Services in Dallas, testing of airborne reconnaissance equipment had not slowed. In January, Cosmonaut Popovich, pilot of *Vostok 4*, was in Cuba to attend the anniversary celebration of the Cuban revolution, where he said, "The world will soon know about the first female cosmonaut." By early June, the Soviet government film industry released a film about a woman training to go into space, and there were rumors that it would happen soon.

28

CHAIKA

On June 16, 1963, not quite a year after the hearings, a twenty-six-year-old woman rode an elevator cage up a gantry adjacent to a launch vehicle. She was wearing a white helmet, and orange nylon-coveralls over a three-layer pressure suit. Where the elevator cage clanked to a stop, eighty feet above the ground, she walked across a short catwalk which led to a white nose cone. There was an opening in that nose cone aligned with the open hatch of the spacecraft it covered. White-coated technicians helped her remove the protective covers from her boots and then helped her through the opening and the open hatch. She had to climb in over the couch on her back, sliding down until the soles of her boots were touching the footrests. They helped her tighten the seat harness, and when she was comfortable, they slammed the hatch shut. She could hear the technicians sealing it by rotating screws into the thirty threaded sockets around its circumference. Slowly, her eyes adjusted to the dim artificial light. It was like the inside of a bathysphere. To her right, along the sloping inner wall, was radio equipment and a control handle for the spacecraft's attitude. Directly before her and just above her head was a recessed porthole. Beneath the porthole was an instrument panel of gauges which would indicate cabin pressure and levels of carbon dioxide and radiation. There was a chronometer with which she could synchronize her activities with those of the ground controllers. At the center of the instrument panel was a rotating globe that would indicate her position over the earth. There were reset switches to its left.

The woman's name was Valentina Tereshkova, and the launch vehicle and gantry were in the eastern Soviet Union, a few miles beyond the outskirts of the city of Tyuratam. Tereshkova knew that Titov had become severely disoriented and nauseous during his flight, so much so that ground controllers had considered cutting the flight short. She knew also that two of the three unmanned *Vostoks* had problems so serious that, had cosmonauts been aboard, they would have been killed. There were rumors of other difficulties. Now, though, was not the time to think of such things. Now, she watched the instrument panel and listened to the steady hiss of cabin pressurization and the voices of launch control as they came in over the headset inside her helmet. She reported readings from the gauges. Soon the voice in her headset told her that the cable mast was about to be disconnected, and she could feel a vibration and shudder through the whole capsule. Then she felt a low distant throbbing like a great heart. Fuel valves had opened, and enormous turbines were churning tons of supercooled liquid oxygen into the great combustion chambers. The missile, now sleeping beneath her, was about to wake.

If Jerrie and Jerri and Bernice and the rest could have spent a day at the Soviets' launch complex in Tyuratam, they might have thought they'd gone through Alice's looking glass. It was like a photographic negative, where another Jerrie Cobb found another head of state who not only shared her dream of a woman in space, but was in a position to have it realized. The Soviet space program was also much shaped by a struggle between politicians and engineering visionaries. But if anything, the struggle was even clearer, because it was between one politician and one engineer.

The West knew the politician all too well. He was Khrushchev. But it knew of the engineer only through reports from the Soviet news agency, Tass, which referred to him as the "Chief Designer." In fact, the chief designer was a trim man in his mid fifties, with handsome features and wavy salt-and-pepper hair. His name was Sergei Korolev. During the Stalinist purge of the thirties Korolev lived for eight years in a gulag, and spent several months of them working in the Kolyma gold mine above the Arctic Circle. Somehow he survived, and for much of the war he worked in another gulag reserved especially for engineers. Afterward, his brilliance gained him his own design bureau and a prominent position in the Soviet missile program. Korolev and his engineers had altered the German V-2 to accommodate a larger and far more powerful engine. Korolev wanted to use it to build a civilization in space. In 1956, when Khrushchev was visiting the missile center in Kalininigrad, Korolev had asked for permission to use an ICBM to launch an earth satellite. Khrushchev agreed, and because Korolev had asked permission, the world entered the Space Age.

Korolev advocated what he called a reasonable approach to space flight, quite like that suggested by von Braun. It would be slow but steady, a program of progressively longer flights and careful reevaluation after each. In the spring of 1961, a year after Gagarin's single orbit, Korolev wanted to launch a manned flight that would last three days. But for Khrushchev the space program was the means by which he would demonstrate the technological superiority of the Soviet Union to the West. And he meant to demonstrate it not only to other engineers, but to the average human being, who was still astonished that other human beings could soar among the stars. So Khrushchev told Korolev he could have his three-day flight, but during that flight he would also have to launch a second cosmonaut, so that two *Vostoks* would be orbiting the earth at the same time. Korolev knew that such a feat would have little or no real scientific or engineering import, and at the time it had little to do with the race to the moon. Korolev knew that he might disobey, but then Khrushchev would simply dismiss him and replace him, in all probability with Vladimir Chelomei, the man who had commanded the engineers' gulag.

There is a Russian folktale about an uneasy relation between a wolf and a fox. The wolf tells the fox that he will eat him unless the fox finds food for him; so the fox finds food for the wolf. The folktale ends happily for the fox, who tricks the wolf in such a way as to cause him to be killed by a farmer. Korolev knew the folktale of the wolf and the fox, and he also knew it was only a folktale.

So he did it. In August 1961 the double flight of *Vostok 3* and *Vostok 4* was judged a success. Afterward, Korolev wanted a flight that would last an entire week, a proper test of man and machine for the time it would take to get to the moon and return. Khrushchev allowed Korolev the week-long flight, but demanded that while that cosmonaut was in orbit, Korolev launch a second spacecraft, this time carrying a woman. Korolev knew that keeping two spacecraft in orbit simultaneously tended to overtax the launch crews and tracking equipment, but it had been accomplished with *Vostoks 3* and *4*. They could do it again. As for the woman, that was another matter. No woman was in training. The practice of the Soviet program, like that of NASA, was to use military test pilots. And as in the United States, all military test pilots were male. All these concerns Khrushchev dismissed. He made it clear that although the woman did not need to be a pilot, she was to be both a Communist and a worker, a clear demonstration to the world that socialist society treats all its people as equals.*

At any rate, in early 1962 the Soviets earnestly set to work finding the first woman to fly into space. They did not ask for volunteers. There was no need to. After Gagarin's flight, the letters had come from every village, every collective, every apartment building in Moscow. The space agency had file cabinets and file cabinets, whole rooms full of these letters. Nobody ever seriously thought they'd be read. But in January 1962 administrators and managers were sent to read them and weed out the unqualified, and those whose political leanings were suspect. After a month of this, five names remained. One name was Valentina Tereshkova. She was twenty-four and single. She had been born on a collective farm near Yaroslavl in 1937. Her father was killed fighting Germans when she was six. After high school she began work as a

*It seemed that Khrushchev regarded American astronauts as having been born into wealth and privilege. This was a rather interesting misperception. Gus Grissom had earned his engineering degree on the GI Bill, and Gordon Cooper's degree in aeronautical engineering had been paid for by the Air Force. In fact, it would not be unfair to say that three or four of the seven Mercury astronauts had used the military as the ticket out of whatever miserable fate would have otherwise been apportioned them. Khrushchev might have been rather surprised to learn that one American woman, who as a child had dressed in burlap and survived by eating pine nuts she found on the ground, had already flown an F-86 Sabre at Mach 1, had taken a Northrop T-38 to 50,000 feet, and had seemed to have made her way into every cockpit *except* that of the Mercury spacecraft.

spindler in a textile factory. She had been secretary of the Communist Youth League. Her only technical qualification was her experience as a parachute jumper, but it was the only one necessary.

In March she received a telegram asking her to report to Lieutenant-General Kamanin in Moscow. In the ensuing months she went through the entire training routine—centrifuge, isolation, flights in a jet flying a parabolic course that simulated weightlessness for a few seconds, even parachute jumps in a pressure suit. The Vostok launched first would carry a cosmonaut named Valery Bykovsky. The second would carry Tereshkova. She would make at least seventeen orbits, spending one day in space. If she showed no ill effects, the flight might be extended. On June 14, Tereshkova watched with other cosmonauts, flight engineers, and Korolev as the preparations for the launch of *Vostok 5* began. Held within the gantry and four girder-like support arms, the Semyorka launch vehicle looked almost delicate; its four conical strap-on boosters gave it the silhouette of a fluted bottle. There was a countdown. The smoke came from beneath, followed by an increasing thunder. The four support arms opened. As Tereshkova watched, the arms slowly swung back, the missile they had held came free, and rose. With ever-increasing speed, Lieutenant Colonel Valery Bykovsky was rising into space.

Two days later, Bykovsky was still in orbit, and Tereshkova was lying on her back inside another spacecraft atop another missile. During the months of her training, Tereshkova had become a friend of Korolev, enough that she understood his dream of a civilization in space. When the chief designer came on the radio link to say, "It's a pity the spacecraft has room only for one," she heard in his voice both paternal concern and undisguised envy. She smiled and said, "Don't worry, Sergei, someday you and I will go to Mars." The countdown started—*Fourteen . . . thirteen . . . twelve . . .* —and she began deliberately to breathe deeply and steadily. She watched the gauges—*Tri . . . Dva . . . Odin . . . Zashiganiye*. Ignition.

She could feel a steady vibration as the engines began, then a series of metallic clanks—the release of the support arms. Beneath the shroud and inside the capsule, Tereshkova could hear a growing thunder and feel a push into the seat. She could feel the push increasing and knew the Semyorka was now free of the supports, climbing into the sky and gathering speed. Thirty seconds later the vibrations had not diminished, but there was quiet, and she knew she was actually outpacing sound itself. The pressure on her lessened, and she heard a distant metallic echoing. She knew the strap-on boosters, now emptied, had separated and fallen away. Now the core stage was firing, and she was pushed again into the seat, harder this time. There was sudden sunlight in the capsule—the aerodynamic shroud was gone. She was already far above any sensible atmosphere, and although it was difficult to move her head, through the porthole above her she could see the sky outside. It was dark blue shading to black. The core stage was still increasing speed and Tereshkova was still being pushed back into her seat. After two minutes the core stage shut down and she was again pushed forward against the straps. Again there was the metallic echoing, now much closer, then sudden quiet, the push back into the seat and a nearer, louder thunder—the ignition of the final stage. It fired for six minutes, pushing her back with a steady, but gentler force. Then it too shut down and dropped away, and she was pushed gently forward. There was quiet again. She realized that she was no longer lying on the couch so much as floating above it, held in place only by her seat restraints. Valentina Vladimirovna Tereshkova was in orbit around the earth, passing eastward over the Soviet Union at an altitude of 110 miles and a speed of 18,000 miles per hour.

The Vostok was made of two components. One was a pressurized sphere some eight feet in diameter. This was the capsule in which the cosmonaut lived, and the only part that would return to earth. Attached to the base of the capsule was the other component—an instrument section, a compact assembly of fuel canisters and fuel lines, all surrounding a braking rocket. Korolev and his designers had only moderate confidence in that rocket. It was liquid-fueled, and there was no guarantee that it would work after several days in the extreme temperatures of the vacuum. The spacecraft had been placed in an orbit whose lowest point skimmed the upper atmosphere just enough to slow it a little bit on each pass. It would, they said, "decay naturally" in ten days. And so in the capsule was enough food, water, and oxygen to last exactly that long.

Her trajectory took her near *Vostok 5*—fewer than five miles away at one point. She and Bykovsky established short-wave radio communications. Through a porthole she could see the spacecraft of her fellow cosmonaut, a bright star over the curved edge of the blue earth. She was flying eastward, and through another porthole she could see the sun setting behind her. Soon she was over the middle of the Pacific, passing into night. And before long she made a telecast. If Tereshkova had any fears of the braking rocket misfiring, they were not in evidence. It was obvious to anyone watching that she was healthy and composed. As she flew over many countries she sent messages: "Warm greetings from space to the glorious Leninist Young Communist League which reared me." "All that is good in me I owe to our Communist party and to the Young Communist League." And so on.

The Russians had a tradition which called for the cosmonaut to name his craft. Tereshkova called hers *Chaika*—"Seagull." Whenever anyone heard this name they smiled. Somehow this name seemed perfect. The graceful white bird sounding its proud and lonely cry above the waves. *Chaika*.

In Red Square in Moscow that Sunday there was a fine, drizzling rain that made the cobblestones shine. But it was mild weather, and so the sidewalks were crowded with people walking. A voice from the loudspeakers set up on many squares announced, "A woman has traveled into space." People stopped so they could hear better, and a young man said to his friends, "Listen, listen to that." An old woman spoke, as though to herself, "Now it has actually happened." That night Tereshkova's spacecraft was sighted in the skies over Paris. Even in America, anyone could go outside after supper, look up at the summer sky, see a point of light slowly tracing an arrow-straight path through the constellations, and know that it was the reflection of sunlight off *Chaika*.

On the second day of the flight Tereshkova did not answer the first calls from the ground, and some of the engineers feared that she was dead. Then they heard her voice, a bit hoarse, and realized that she had been sleeping soundly and was merely very comfortable. That evening before her planned sleep period Korolev came onto the radio link. Tereshkova said, "I fell asleep for some time contrary to schedule. Excuse me. I shall do better." Korolev told her that it did not matter that she had fallen asleep, and asked her to adjust the cabin temperature. Then he asked her if she wished to end the flight, and she answered that on the contrary she wished to fulfill the entire program. There was a pause, and he said, "Let us not decide that now. The morning is wiser than the night. Sleep quietly." She did sleep quietly; telemetry showed that throughout the rest period her heart was beating at fifty-two to fifty-four beats per minute. In the morning Korolev extended the flight from seventeen orbits to forty-eight. Tereshkova operated the manual orientation system, took photographs of the moon and stars and especially of the Earth's twilight corona—the layers of the atmosphere lit by the sun. She supervised several biological experiments involving insects and seeds. She diagnosed her physical condition, read observations into the voice-activated tape recorder, exercised, and ate. And she made more broadcasts.

Finally, as Tereshkova neared the end of her forty-eighth orbit, the engineers were given the order to begin the reentry sequence, and Korolev came on the radio link and informed the cosmonaut. She replied, "I understand. I am ready for the descent." She pulled the restraining harness tighter. If all went as planned, the couch would become her ejection seat. The Vostok landed not in the ocean as did the Mercury capsule, but on the earth. And because it would hit hard, the cosmonaut did not land with it. So the couch was on rails, and there were two solid rockets at its base. At an altitude of about 20,000 feet the hatch over the cosmonaut's head would be jettisoned, the two solid rockets would fire, and the couch—with the cosmonaut strapped to it—would be shot from the capsule. As the couch fell, it would automatically deploy a small stabilizing parachute. A few moments later the cosmonaut, assuming she were conscious, would release her harness and kick herself free of the seat. And her own parachute would open. If all went well. On the first test of this system in the spring of 1961, a Vostok with a man inside was carried aloft in the cargo section of a transport plane. The plane's cargo hatch opened, the Vostok was dropped, the man tried to eject and his head struck the hatch. Later, the recovery crew found his body in the frozen mud below.

But for Tereshkova, reentry was still minutes away. Now, headed east over the South Atlantic, the automatic orientation system switched on, and thrusters turned the Vostok so that it was flying engine nozzle first. Before long she was over Africa. As the braking rocket fired, she felt the weight she had not felt in three days, and again the distant clanking sounds. The instrument module had completed its work, and now was uncoupling itself. She reset the clock on the instrument panel, then lowered the visor on her helmet and locked it. Now the capsule was falling earthward. As it touched the upper fringes of the atmosphere she could feel it turn, finding its center of gravity. She felt a pressure thrusting her into her seat, and the capsule began to buffet. Ten minutes to ejection. Slowly the push into the seat increased from five g's to six. It became difficult to move her head. At seven and eight g's, the edges of the porthole glowed orange, and she was pressed into the seat with such force that breathing itself was a conscious effort. There was a steady roar through the capsule walls, and flames visible through the porthole were illuminating the inside of the capsule. Now the air around the capsule was ionized, and no radio signal could get in or out. Five minutes to ejection. Gradually the capsule was slowed. The roar subsided and the flames were gone. The drogue chute opened, and she felt a sudden jolt. She watched the sweep hand on the chronometer—*ten seconds to ejection . . . eight . . . six . . . three . . . two . . .* A red light flashed on the indicator panel and the hatch over her head was blown off. There was sudden daylight inside the capsule. Even in her helmet and closed visor she could hear air rushing through the cabin. She braced herself and counted. The two solid rockets fired, and she was shot out of the capsule into open air. She was tumbling, the ground and sky spinning around her. Then a small parachute shot from the seat, opened, and jerked her upright. She released her harness, and kicked herself free of the seat. At 1,200 feet she felt a jolt as her own parachute opened. She was hanging beneath it, swinging a little, drifting high over the ground. From here on it was just another late morning parachute jump, like dozens of others. Still drifting, she opened her visor—the air was cool on her face. She was breathing hard, but everything around her was quiet, still. She could hear a dog barking somewhere below. And she realized she could smell the black sun-warmed earth of the steppe.

Tereshkova had reentered in the Kazakh Uplands about three hundred miles northeast of Karaganda. It was a country of low hills covered with turf grasses, and here and there were small herds of sheep. Nearby, some men were building a bridge. They saw the descending parachute, and the tiny orange-suited figure suspended beneath it. They watched it sink behind the low hills, and saw the parachute lose its shape and sink, too. When they ran over those hills they saw a woman standing in a field, unbuckling her parachute harness. On the ground near her feet was a space helmet. They knew who it was. Even here in the provinces, they had heard of the woman named Seagull, who flew among the stars.

She had to take off her pressure suit, open the container also ejected from the Vostok and change into a track suit that was inside. Then she had to collect the pressure suit and parachute and carry them to the Vostok, which had landed twelve hundred feet away. It was sitting there in the high grass, a great sphere blackened like charcoal, parachute straps draped over it, the parachute itself spread over the ground nearby. The men approached, and helped her carry the ejection seat. They welcomed Tereshkova as befit a national hero. They gave her the traditional Russian welcome of a meal of bread and salt. She asked them to drive her to the nearest village with a telephone. There, she put a call through to the Kremlin. Then she returned to the field to await the rescue aircraft.

When Khrushchev had talked with her on the phone he said, "Your voice rings as though you have come from a party." The man was exultant, and he had reason. By any measure Tereshkova's flight was a spectacular success—nearly three days and forty-eight orbits. In a single flight she had logged more time in space than had all of NASA's astronauts over the course of the entire Mercury program. Tereshkova was flown to a town on the Volga where cosmonauts recuperated after flights. There she met Bykovsky, who had returned to earth an hour after her. On a balcony before a cheering crowd, they embraced. Two days later both went to Moscow. In Red Square there were thousands of cheering people. Khrushchev embraced her, and she and Khrushchev and Bykovsky and the other cosmonauts stood on a reviewing stand. There were flowers everywhere; it seemed that everyone wanted to give her a bouquet. And Khrushchev, the man who looked like an ogre to much of the foreign press and even to his own people, now looked like nothing so much as a kindly uncle. He spoke of the peaceful conquest of outer space and of an upcoming meeting among the

U.S., the Soviet Union, and Britain to discuss a nuclear test ban. But he could not seem to concentrate on the prepared text. He kept stopping himself to look at Tereshkova and smile. Finally he put the text away, looked at Tereshkova and grinned and shook his finger at the crowd and said, "*Here is your weaker sex. . . .*"

The Russians had done it first. Back in the United States Jane Hart was interviewed, and said she wasn't a bit surprised that the Russians were first and that NASA didn't seem to care. She said the Russians could have landed the whole Leningrad Symphony Orchestra on the moon and returned it, and NASA probably still wouldn't let a woman fly. Jerri Sloan in Dallas just simmered—it was a dull pain suddenly made sharp again. In Oklahoma City, a few reporters remembered something about American women astronauts, and somebody checked the files, found a telephone number at Aero Commander, and dialed it. They might as well have asked Cassandra how she felt on the morning after the fulfillment of her prophecy. But Jerrie was polite, and she spoke to them as a midwesterner speaks to strangers: "It's a shame that since we are eventually going to put a woman into space, we didn't go ahead and do it first."

On Capitol Hill again there was talk of women astronauts. The Senate Aeronautical and Space Sciences Committee met with Robert Seamans, associate administrator of NASA. Seamans defended the policy of restricting astronauts to test pilots with engineering degrees or the equivalent. He acknowledged that the policy effectively eliminated women, but said that it was not established for that reason. Senator Clinton P. Anderson, chairman of the committee, said perhaps we are being a little "stiff-necked in our attitude," and Senator Stuart Symington suggested a compromise—including a single woman among the astronaut corps.

Jerrie allowed herself a little hope again, and told the other women, "Maybe we'll get some action." In the year since the hearings Wally Funk had been tested on a centrifuge and in a Marine high-altitude chamber, and although Jerri Sloan had said to hell with it after the tests at Pensacola were canceled, she was still exercising and watching her diet. Some of the press were still interested; usually they'd run a story as a sidebar. *Life* magazine did an article on Tereshkova, showing her riding a bicycle in a park and getting her hair done. She seemed to be a real "girl-next-door," that is, of course, if the girl next door happened to be former chapter secretary of the Young Communist League. And *Life* followed that story with two pages on the women tested at Lovelace, including photographs. The title of those two pages was, "The U.S. Team Is Still Warming Up the Bench."

In the United States during the next weeks and months there were rumors that Tereshkova had been locked into the capsule screaming and kicking, that she had been hysterical for most of the flight. Most serious observers of the Soviet space program did not credence these stories, and neither did NASA officials. Still, NASA would deride the effort as a mere "stunt," in part because Tereshkova was a passive passenger, not a pilot. In fact, America's most recent space flight—it was Gordon Cooper's—had done more to vindicate the argument for the astronaut as pilot. The spacecraft's automatic systems failed completely, and Cooper was forced to reenter manually.

So for a few weeks in the summer of 1963 the talk about the *role of the astronaut* started all over again. This time, the women had some well-placed allies. In an editorial in *Life*, Clare Booth Luce argued that Communism assumed that men and women are inherently equal, and made good on that assumption with an especially aggressive sort of equal-opportunity program, of which Tereshkova was merely a recent result. Luce had the numbers to prove it, and they were impressive. On the eve of the revolution in 1917 there were 300 women engineers; by 1961 that number had increased more than a thousandfold to 379,000—almost a third of all the engineers in the Soviet Union. In 1961, 53 percent of all professionals in the Soviet Union were women, women composed 26 percent of the Supreme Soviet, and most impressive, the Soviet Union by 1962 could boast that nearly *three quarters* of its physicians were women, an astonishing total of 332,400. The United States, in the same year, had a miserable 14,000. And, Luce said, if Tereshkova was a political tool—well, so what? She was a political tool they had earned the right to use. We should be so lucky.

The women had another friend in Senator Ernest Gruening of Alaska, a vocal proponent of a woman-in-space program. On June 27, only eight days after Tereshkova's return, he addressed the

Senate. He read from Booth's editorial. And he said that the Soviet Union's sending a woman into space was a greater defeat for the United States than even Sputnik had been. His point was that whatever the propaganda value, there was no getting around the simple fact that the Russians had entrusted a cosmonaut mission to someone he called "a 26-year-old girl." Clearly, he said, the Soviet Union believed that women possessed the same abilities as men. And just as clearly, the Soviet Union was willing to act upon that belief. And it didn't escape the notice of the women—with their average of 4,500 hours, even if they were only in props and transports—that Tereshkova was not even a pilot.

29

INNER SPACE

The events of October 1962 seemed to awaken the United States and the Soviet Union to the dangers of nuclear brinkmanship. On April 5, 1963, a "hot line" agreement was reached between the two nations, and on August 5, they concluded a partial test-ban treaty, outlawing nuclear tests in the atmosphere, underwater and in outer space. Kennedy considered it one of the most significant accomplishments of his administration. The president was widely perceived as an advocate of NASA's plans; putting a man on the moon had been his challenge, and the newspapers would not let him forget it. Privately though, he had remained troubled by the costs. To the surprise of many, in an address on September 20 to the United Nations' General Assembly, he proposed that the United States and the Soviet Union forgo the race, and send a joint expedition to the lunar surface.

He would not live to see the idea fulfilled, and America would not remember it. At the funeral in November a young White House aide named Daniel Patrick Moynihan heard a friend standing near him say, "We'll never be young and happy again." Moynihan said, "No. We'll be happy again. But we'll never be young again." The comment was prescient. In the summer of 1964, a year after Martin Luther King Jr. had led a march for civil rights in Washington, a white policeman shot and killed a black youth in Harlem. The tragedy ignited five nights of race riots. Before the year was out, American warplanes had attacked North Vietnamese torpedo boat bases, and the newspaper headlines were about places like Hanoi and the Gulf of Tonkin.

During those months and years the women tested at Lovelace worked and planted gardens in the backyard and in one way or another kept flying. From their homes they watched the space program on television. For a few months the letters from Jerrie still came. They'd still begin, "Dear Fellow Lady Astronaut Trainees," but after a while that wishful salutation would sound less and less real. Finally the letters ceased.

It is difficult to say exactly when the cause was lost. Some might say that it was the day in June 1962 when Cobb and Hart walked out of Room 356 and down the steps of the Cannon Building. Probably though, the women lost the battle long before. They had already lost it when Kennedy called for a man on the moon in ten years. They had already lost it when, four years earlier, Eisenhower ordered that astronaut candidates be military test pilots. They had already lost it when von Braun's civilization among the stars was sacrificed to the exigencies of Cold War politics. In fact, they had lost eighteen years earlier, on December 20, 1944, the day Ann Baumgartner and Jean Hixson and more than one thousand other women were disqualified as military test pilots, the day the WASP were deactivated.

But the thirteen women tested at Lovelace Clinic in 1961 were disinclined to backward glances. They went on with their lives and careers, and some began new ones. Gene Nora Stumbough would say that the canceled tests at Pensacola were the best thing ever to happen to her. In 1962, when Jerrie Cobb was the only woman in the United States flying for a manufacturer, Gene Nora found a position as a sales demonstration pilot with Beech Aircraft Corporation, and met the man she would marry. Jerri Sloan married Joe Truhill in 1966, and Air Services continued to test infrared systems. By then the tests at Lovelace had become a fading and bittersweet memory.

In 1965 Jerrie Cobb began a second career as a humanitarian pilot working with doctors, missionaries, anthropologists, and linguists in the Amazon basin. She ferried medicines and supplies, and flew critically ill or injured people to clinics. She called her twin-engine amphibian *Little Bird*, and locally she would become known as the "Flying Angel of Amazonia." To friends who asked her whether she thought much about her attempts to become an astronaut, Jerrie would respond that she had found much greater challenges.

EPILOGUE

Nancy Harkness Love died of cancer in 1976. She was sixty-two. Among her personal effects was a box containing a scrapbook, a handwritten list of names of women with commercial pilot's licenses in 1940, and news clippings and photographs of each of the women who died under her command.

In June 1963, a year after the end of Jerrie Cobb's tenure in the position, NASA administrator James Webb named Jackie Cochran an official consultant to NASA. She would retain the position until 1969. Meanwhile, she would continue to fly faster planes. In 1964, when she was at least fifty-four and perhaps as old as fifty-eight, she piloted a Lockheed F-104G Starfighter at twice the speed of sound. During those years she would still read the morning paper, and if she were particularly bothered by some development in aerospace, she exercised her right as an informed citizen and simply called the White House. On one occasion the White House operator informed her that the president (it was Johnson then) was in a meeting. She told the operator to tell him to call her. Later that day the phone rang, Floyd answered, and was told that the president of the United States was returning his wife's call. He held the phone out to Jackie, and Jackie didn't move. Instead she told Floyd to tell Johnson that she was washing her hair and that he could damn well call back later. Floyd told the operator to tell Johnson he could call back later. And Johnson called back later.

In 1971 Jackie suffered seizures, and was diagnosed with heart and circulatory difficulties. An operation by Randy Lovelace and a pacemaker alleviated the condition somewhat, but the chance of another seizure and changes in air pressure at altitude disallowed her from flying even light aircraft. In the early 1970s Floyd would make an uncharacteristically ill-advised business decision that would require them to sell the ranch to a condominium developer. Floyd had long suffered from arthritis, and by 1974 it had worsened to the point that he was mostly bedridden; he would allow only Chuck Yeager or a friend named Yvonne Smith to move him from bed to his wheelchair. In 1976, Floyd died in Jackie's arms. He was eighty-four.

Shortly afterward, while helping Jackie with personal papers, Yeager discovered the envelope containing the detective's findings regarding Jackie's birth, and asked her if there was any reason to open it or even to keep it. She said nothing. And he burned it.

In her last years, Jackie's friends said, she lost much of her spirit, in part because she could no longer fly, in part because Floyd was gone. Soon her own health declined further. She had heart and kidney failure, and often was in such terrific pain that she had to sleep sitting upright. Many of her friends could not bear to visit her, and she would say that only Yeager could make her laugh. She began to speak of death, and told several friends that she wanted to be buried with the doll she had won in the lottery as a child, and with a sword given her in 1976 by Air Force Academy cadets. Of the latter request she would explain, "I want to take this sword with me if I go to Heaven, and well, if I go to that other place, I'll be able to fight my way out."

Jackie Cochran died in August 1980. Her obituary in *The New York Times* said her age had been listed variously as seventy and seventy-four. The sword was returned to the Air Force Academy, where it remains today; the doll was buried with her.

In the spring of 1990, as part of the commemoration of the fiftieth anniversary of Russian involvement in World War II, thirty WASP met in Moscow with members of the Night Witches. The women marched together in the May Day parade in Red Square.

In several interviews, Valentina Tereshkova insisted that the nearly two years she spent in training would not have been necessary for a flight that was mere propaganda; if the purpose had been merely to launch a woman first, the Soviets could have made any healthy woman a passenger aboard a spacecraft piloted by another cosmonaut. Nonetheless, Tereshkova did not fly again in space, and at least for a few years the Soviet Union abandoned its use of women cosmonauts. In 1968 cosmonaut Yuri Gagarin died in a plane crash. His remains were cremated and placed in an urn, which was put in a niche in the Kremlin Wall. There was a state funeral, and four persons were given the honor of carrying the catafalque on which the urn rested. All were cosmonauts, and one was Tereshkova. As more information on the Soviet space program became available to Western sources, there arose suspicion that the circumstances surrounding Tereshkova's flight were stranger than anyone had thought. The five women candidates were single, and the only single cosmonaut, Andrian Nikolayev, had been charged with their training. He and Tereshkova became involved well before her flight, perhaps before she was even chosen as a candidate. In November, four months after her flight, she and Nikolayev were married. Amidst much fanfare, Khrushchev himself presented the bride. A few months after the wedding Tereshkova gave birth to a daughter they named Yelena.

Responding to concerns from many groups and individuals, NASA in the mid-1970s began to recruit women and minorities; the results of its efforts were evident in the eighth astronaut group, selected in 1978. Four of the thirty-five were minorities, and six were women, among them Sally Ride and Shannon Lucid.* By the 1980s one might have thought competition in outer space was over. The U.S. had won the race to the moon, while the official Soviet line was that there had never really been a race. But in 1982 the Soviets learned that NASA was preparing to launch a woman. On August 19, 1982, the Soviets preempted Sally Ride's flight, launching the second woman in space, Svetlana Savitskaya, along with two male cosmonauts aboard a Soyuz spacecraft. When they docked with the Salyut Space Station, Savitskaya was greeted by the cosmonaut-in-residence with flowers, an apron and a joking invitation to do the cooking. She is said to have replied that meal preparation is the responsibility of the host cosmonaut.

At present, Jerri Sloan Truhill is retired from professional flying. Bernice Steadman continued to own and operate her flying service for many years. More recently, she has worked for the International Women's Air and Space Museum, serving both as president and chairman of the board. Sarah Gorelick Ratley never returned to AT&T, and now works as an accountant in Kansas City. In 1967 Jean Hixson began work at the flight simulation techniques branch of the Air Force Reserves at Wright-Patterson Air Force Base. She was chair of the WASP reunion in 1982, during which she piloted a B-25. She retired a full colonel from the Air Force Reserves in that same year, having been awarded the Meritorious Service Medal. The following year she concluded a thirty-year career as an educator, and retired from the Akron Ohio school system. Shortly thereafter she died of cancer, at age sixty-two. Jane Hart has been retired for many years, and now resides in the Virgin Islands. Gene Nora Stumbough Jessen and her husband moved to Boise, Idaho, where they started a Beech Aircraft dealership, and where she resides today. In 1974 Wally Funk became one of the first female Air Safety Investigators with the National Transportation Safety Board. She works now as an aviation safety counselor,

*In total, 268 U.S. astronauts have been selected in the sixteen groups from 1959 through 1996; there are 98 astronauts and 35 candidates currently in the program; 104 astronauts have retired, resigned or been reassigned; and 25 are dead.

lecturing all over the United States. Kay Cagle returned to flying, and is currently certified as an instructor and a Civil Air Patrol pilot. Rhea Allison Woltman stopped flying professionally shortly after the tests; she now resides with her husband in Colorado, where she works as a parliamentarian. Marion Dietrich continued to write about aviation for the *Oakland Tribune*, and later wrote for both the *San Francisco Chronicle* and *Time* magazine. She died of cancer in 1974, and her sister Jan scattered her ashes from a plane over the Pacific. Irene Leverton and Jan Dietrich, acquaintances before the tests, came to know each other better later, and both worked in California. Jan is presently in ill health. Recently, in a discussion of astronauts and astronaut qualifications, Irene grew quiet for a moment, and said of her friend, "She would have made a good one." Irene still flies part-time for a manufacturing company in Prescott, Arizona, gives check rides, and acts as a safety officer for the local Civil Air Patrol squadron. Recently she has begun to dabble in oil painting, mostly of aircraft.

Jerrie Cobb has spent nearly thirty years flying supplies and medicine into remote areas of Colombia, Venezuela, Brazil, Bolivia, Peru, and Ecuador. She has learned more than sixteen dialects, suffered a bout with malaria, and had encounters with unfriendly guerrilla forces. In 1980 Representative Mickey Edwards of Oklahoma nominated her for the Nobel Peace Prize, writing in his nominating letter, "Jerrie Cobb is a unique individual who has devoted all of her skills and resources to providing health, bringing hope, and creating peace for thousands of men, women and children." In February 1995, Jerrie, with six of the women tested at the Lovelace Clinic in 1961, and several WASP, attended the launch of Eileen Collins.

Eileen Collins's flight in February 1995 was a success. The mission achieved not only a rendezvous with Mir, but the deployment and retrieval of an astronomy satellite, and a spacewalk. In eight days Collins and the crew traveled 2.9 million miles. In May 1997 the women tested at Lovelace were invited to Eileen Collins's second flight, in which she acted as pilot for STS-84, a mission which would dock with Mir. Among its crew members was mission specialist Elena Kondakova, a cosmonaut who had spent 169 days in space aboard the spacecraft "Soyuz TM-17" and Mir.

Besides Eileen Collins, two women are currently qualified to fly the Shuttle. Susan Still was pilot for STS-83 in April 1997 and STS-94 in July 1997, and Pamela Melroy is currently assigned as pilot on STS-92.

On March 5, 1998, NASA named Eileen Collins the first woman commander of an American space flight.

SELECTED BIBLIOGRAPHY

Books and Articles

Atkinson, Joseph D., and Jay M., Shafritz. *The Real Stuff: A History of NASA's Astronaut Recruitment Program*. New York: Praeger Publishers, 1985.

Bell, Elizabeth S. "The Women Fliers: From Aviatrix to Astronaut." In *Heroines of Popular Culture*, ed. Pat Brown. Bowling Green, Ky.: Bowling Green State University Popular Press, 1987.

Bilstein, Roger E. *Orders of Magnitude: A History of the NACA and NASA, 1915–1990*. Washington, D.C.: National Aeronautics and Space Administration, 1989.

Blashfield, Jean F. et al., eds. *Above and Beyond: The Encyclopedia of Aviation and Space Sciences*. Chicago: New Horizons Publishers, Inc., 1967.

Boase, Wendy. *The Sky's the Limit: Women Pioneers in Aviation*. New York: Macmillan, 1979.

Breuer, William, B. *Race to the Moon: America's Duel with the Soviets*. Westport, Conn.: Praeger Press, 1993.

Brooks-Pazmany, Kathleen. *United States Women in Aviation 1919–1929*. Washington, D.C.: Smithsonian Institution Press, 1983.

Cadogan, Mary. *Women with Wings: Female Flyers in Fact and Fiction*. Chicago: Academy Chicago Publishers, 1993.

Carpenter, M. Scott, et. al. *We Seven*. New York: Simon & Schuster, 1962.

Cobb, Jerrie, with Jane Rieker. *Woman into Space: The Jerrie Cobb Story*. Englewood Cliffs, N.J.: Prentice-Hall, 1962.

Cochran, Jacqueline, with Floyd Odlum as Wing Man. *The Stars at Noon*. Boston: Little, Brown & Co., 1954.

———, and Maryann Bucknum Brinley. *Jackie Cochran: An Autobiography*. New York: Bantam Books, 1987.

Cole, Jean Hascall. *Women Pilots of World War II*. Salt Lake City: University of Utah Press, 1992.

"Damp Prelude to Space." *Life*, October 24, 1960, p. 88.

Dempsey, Charles A. *Air Force Aerospace Medical Research Laboratory: Fifty Years of Research on Man in Flight*. Washington, D.C.: U.S. Air Force, 1985.

Dietrich, Marion. "First Woman into Space." *McCall's*, September 1961.

Douglas, Deborah G. *United States Women in Aviation: 1940–1985*. Washington, D.C.: Smithsonian Institution Press, 1991.

Everest, Frank K. and John Geunther. *The Fastest Man Alive*. New York: Pyramid Books, 1959.

Flora, Michael R. "Project Orion: Its Life, Death, and Possible Rebirth." Unpublished essay, 1993.

Glenn, John, ed. *Letters to John Glenn*. Houston: World Book Encyclopedia Science Service, Inc., 1964.

Granger, Byrd Howell. *On Final Approach: The Women Airforce Service Pilots of W.W.II*. Scottsdale, Ariz.: Falconer Publishing Co., 1991.

Hallion, Richard P. *Test Pilots: The Frontiersmen of Flight*. Washington, D.C.: Smithsonian Institution Press, 1988.

Harvey, Brian. *Race Into Space: The Soviet Space Programme*. New York: John Wiley & Sons, 1988.

Hodgman, Ann, and Rudy Djabbaroff. *Skystars: The History of Women in Aviation*. New York: Atheneum, 1981.

Jackie Cochran File, NASA History Office, Washington, D.C.

Jackson, Robert. *Mustang: The Operational Record*. Washington, D.C.: Smithsonian Institution Press, 1992.

Jerrie Cobb File, NASA History Office, Washington, D.C.

Johnson, Nicholas L. *Handbook of Soviet Manned Space Flight*. Vol. 48 of

Science and Technology Series. San Diego: American Astronautical Society, 1980.

Kennedy, Susan Estabrook. *If All We Did Was to Weep at Home: A History of White Working-Class Women in America*. Bloomington: Indiana University Press, 1979.

Knapp, Sally. *New Wings for Women*. New York: Thomas Y. Crowell Company, 1946.

Laboda, Amy. "The Mercury 13." *AOPA Pilot* 7 (February 1997): 61–66.

"A Lady Proves She's Fit for Space Flight." *Life*, August 29, 1960, pp. 73–76.

Link, Mae Mills. *Space Medicine in Project Mercury*. Washington, D.C.: National Aeronautics and Space Administration, 1965.

Lothian, A. *Valentina: First Woman in Space*. Cambridge: Pentland Press, 1993.

Luce, Clare Booth. "A Blue-Eyed Blonde in Orbit." *Life*, June 28, 1963, pp. 28–34.

May, Charles P. *Women in Aeronautics*. New York: Thomas Nelson & Sons, 1962.

Miller, Richard L. *Under the Cloud: The Decades of Nuclear Testing*. New York: The Free Press, 1986.

Moolman, Valerie and the Editors of Time-Life Books. *Women Aloft*. Alexandria, Va.: Time-Life Epic of Flight Series, 1981.

Myles, Bruce. *Night Witches: The Untold Story of Soviet Women in Combat*. London: Mainstream Publishing, 1981.

Nalty, Bernard C., John F. Shiner and George M. Watson. *With Courage: The U.S. Army Air Force in World War II: A Commemorative History*. Washington, D.C.: U.S. Government Printing Office, 1994.

Noggle, Anne. *For God, Country, and the Thrill of It: Women Airforce Service Pilots in World War II*. College Station, Tex.: Texas A & M University Press, 1990.

Oberg, James E. *Red Star in Orbit*. New York: Random House, 1981.

Osur, Alan M. *Blacks in the Army Air Forces During World War II: The Problem of Race Relations*. Washington, D.C.: Office of Air Force History, 1977.

Pebbles, Curtis. *Dark Eagles: A History of Top Secret U.S. Aircraft Programs*. Novato, Calif.: Presidio Press, 1995.

Pennington, Reina. "Wings, Women and War." *Air and Space*. December 1993/January 1994.

Phelps, J. Alfred. *They Had a Dream: The Story of African-American Astronauts*. Novato, Calif.: Presidio Press, 1994.

Popescu, Julian. *Russian Space Exploration*. Oxon, Eng.: Gothard House Publications, 1979.

"Qualifications for Astronauts." Report of the Special Subcommittee on the Selection of Astronauts, Committee on Space and Aeronautics, U.S. House of Representatives. Washington, D.C.: U.S. Government Printing Office, 1962.

Reeves, Richard. *President Kennedy: Profile of Power*. New York: Simon & Schuster, 1993.

Rich, Ben E. and Leo Janos. *Skunk Works: A Personal Memoir of My Years at Lockheed*. New York: Little, Brown and Company, 1994.

Roberts, David. "Men Didn't Have to Prove They Could Fly, but Women Did." *Smithsonian*, August 1994.

Rodman, Selden, ed. *The Poetry of Flight*. New York: Duell, Sloan and Pierce, 1941.

Roosevelt, Eleanor. "Flying is Fun." *Colliers*, April 22, 1943.

Sandler, Stanley. *Segregated Skies: All-Black Combat Squadrons of World War II*. Washington, D.C.: Smithsonian Institution Press, 1992.

Santy, Patricia A. *Choosing the Right Stuff: The Psychological Selection of Astronauts and Cosmonauts*. Westport, Conn.: Praeger, 1994.

Scharr, Adela Reik. *Sisters in the Sky*, vols. 1 and 2. St. Louis: The Patrice Press, 1988.

Shepard, Alan and Deke Slayton with Jay Barbree and Howard Benedict. *Moon Shot: The Inside Story of America's Race to the Moon*. Atlanta: Turner Publishing, Inc., 1994.

Slayton, Donald K. "Deke" with Michael Cassuit. *Deke!—U.S. Manned Space: from Mercury to the Shuttle*. New York: Tom Doherty Associates, 1994.

Smith, Elizabeth Simpson. *Breakthrough: Women in Aviation*. New York: Walker & Co., 1981.

Swenson, Jr., Loyd S., James M. Grimwood and Charles C. Alexander. *This New Ocean: A History of Project Mercury*. Washington, D.C.: National Aeronautics and Space Administration, 1966.

Thomas, Shirley. *Men of Space: Profiles of the Leaders in Space Research, Development, and Exploration*. Vol. 2. New York: Chilton Company, 1961.

Van Wagenen Keil, Sally. *Those Wonderful Women in Their Flying Machines*. New York: Rawson, Wade, Publishers, Inc., 1979.

Verges, Marianne. *On Silver Wings: The Women Airforce Service Pilots of World War II*. New York: Ballantine Books, 1991.

Von Bencke, Matthew J. *The Politics of Space: A History of U.S.-Soviet/ Russian Competition and Cooperation in Space*. Boulder, Colo.: Westview Press, 1997.

Von Braun, Werner. *The Mars Project*. Urbana: University of Illinois Press, 1953.

Williams, Vera S. *WASP: Women Airforce Service Pilots of World War II*. Osceola, Wis.: Motorbooks International, 1994.

Wolfe, Tom. *The Right Stuff*. New York: Bantam Books, 1980.

Yeager, Chuck and Leo Janos. *Yeager: An Autobiography*. New York: Bantam Books, 1985.

Museums and Collections

Eisenhower Museum, Abilene, Kansas

International Women's Air and Space Museum, Centerville, Ohio

NASA Archives, Washington, D.C.

The Ninety-Nines, International Women Pilots, Historical Archives. Oklahoma City, Oklahoma

Smithsonian Institution, Washington, D.C.

Texas Woman's University Libraries, Denton, Texas

INDEX